Building on Knowledge

Developing Expertise, Creativity and Intellectual Capital in the Construction Professions

David Bartholomew

WILEY-BLACKWELL

A John Wiley & Sons, Ltd., Publication

This edition first published 2008
© 2008 Blackwell Publishing Ltd

Blackwell Publishing was acquired by John Wiley & Sons in February 2007.
Blackwell's publishing programme has been merged with Wiley's global Scientific,
Technical, and Medical business to form Wiley-Blackwell.

Registered office
John Wiley & Sons Ltd, The Atrium, Southern Gate, Chichester, West Sussex,
PO19 8SQ, United Kingdom

Editorial offices
9600 Garsington Road, Oxford, OX4 2DQ, United Kingdom
2121 State Avenue, Ames, Iowa 50014-8300, USA

For details of our global editorial offices, for customer services and for information
about how to apply for permission to reuse the copyright material in this book
please see our website at www.wiley.com/wiley-blackwell.

ISBN: 978-1-4051-4709-5

Library of Congress Cataloging-in-Publication Data is available

A catalogue record for this book is available from the British Library.

Set in 10 on 12.5 pt Avenir by SNP Best-set Typesetter Ltd., Hong Kong
Printed in Singapore by Fabulous Printers Pte Ltd

1 2008

Contents

Preface

My interest in knowledge began over 25 years ago when I was responsible for directing the UK's national programme of research on solar energy. As results started to roll in it became clear that, though some of the ideas belonged to the future, others deserved to be taken up immediately: they could make buildings cheaper to run and nicer to live and work in, without costing anything. Only the designs would need to change. We published the research reports, but we soon found that hardly anybody in the construction industry reads research reports. How could we get these new ideas across to them? How, indeed, did knowledge in general flow from research into practice?

When I looked into past innovations, I found that it could take up to 20 years for new ideas to spread throughout a whole industry (and construction was not uniquely slow). I did not want to wait that long. Even after new ideas reached one part of a company they often took a long time to become common knowledge. How could I speed up the process? I became fascinated by knowledge and how it flows around, between organisations and inside them. I discovered a lot about how ideas emerge and practical know-how develops; that some kinds of knowledge can be communicated easily in writing, some only with difficulty, and some not at all; the importance of tacit knowledge (but I had not read Polanyi's book, so I did not know there was a name for it) and the special magic of face-to-face knowledge transfer; and many other things. And this mysterious stuff called knowledge has been part of my professional life ever since.

This book was inspired by a series of three research projects about knowledge in construction – principally in architectural practices, engineering consultancies and client organisations – which I initiated and led between 1998 and 2005. They focused respectively on the use of IT to make information more easily and quickly accessible, on learning from project experience, and on sharing knowledge within organisations – between them, all the main processes involved in what we have come to call 'knowledge management'. We wanted to discover how organisations can improve quality, avoid endlessly reinventing wheels and repeating mistakes, reduce risk, become more creative, make working life more enjoyable, and improve their bottom lines.

Looking back, four things stand out: that understanding of knowledge management has developed enormously in the past decade, thanks largely to the accumulation of evidence from practice; that the same principles, tools and techniques have emerged as central to success in all kinds of organisation (including the principle that *implementation* needs to be tailored sensitively to the organisational context); that, despite these commonalities, professional services organisations have features that pose special problems for managing knowledge; and that, despite the many books – some of them excellent – that have been published on the subject, business leaders *still* find it hard to discover what to do. This is an attempt to fill some of the gap.

The practitioners I have worked with tell me they find the existing literature variously too academic, too didactic, too specialised, too abstract, too much concerned with the alien world of big corporations, and simply too extensive. Much of it also makes knowledge management sound alarmingly complicated. Software vendors, by contrast, claim that knowledge management is simply a matter of buying their (hugely expensive) 'solutions'. This is seductive, but it has long since been exposed as misleading at best and often a quick road to disappointment. Well-informed boards are now wisely sceptical.

This is, therefore, a consciously *non*-academic, *un*didactic, wide-ranging and relatively short book, which looks at knowledge management from a practical and specifically professional services perspective. There is, for example, no chapter reviewing the history of research or thought in knowledge management, and I have not tried to relate the various tools and techniques to academic theories. And although, like all books on knowledge management, this one has roots in a wide variety of ideas from management guru Peter Drucker (who coined the term 'knowledge worker' around 50 years ago) onwards, I have cited authorities for only a few of them. Many appear anyway to have arisen independently in several places. That is not surprising in a field where the best evidence for most propositions is simply personal experience. To adapt the inscription on Christopher Wren's tomb in St Paul's Cathedral: *lector, si argumentum requiris, circumspice.*[1]

It is not simply a collection of recipes, either. The trouble with recipes alone is that they give the cook no help when the meal needs to be adapted for different ingredients, equipment or tastes, because

---

[1] Reader, if you seek the justification, look around you.

they fail to explain *why* certain things work and others do not. Since the basic recipes of knowledge management always have to be adapted to suit organisations' individual circumstances, some understanding of underlying causalities is vital. This I have tried to provide. Nevertheless, well-tried recipes are invaluable as a starting point, so they are here too.

I have focused particularly on the needs of consultancy practices and repeat clients – architects, engineers, surveyors, and clients in sectors such as education, health, government, retail and the utilities – but much of the book should be equally relevant to the professional aspects of contracting. It does not, though, address transactional aspects of construction such as e-tendering and the handling of project documentation, which are essentially exercises in managing data and information, not knowledge.

It is probably too much to hope that people in other professional services will get past the title, but if they do look further they will find much that is relevant to them, too. Most professional services organisations share characteristics that differentiate them from the large manufacturing and commercial corporations which dominate the knowledge management literature. Most of them are small; their work is usually project based rather than process based; their staff are relatively homogeneous in educational background and function; management structures are flat; and, often, the managers are also the owners, and carry on their professional work alongside management. These all have important consequences for managing knowledge, which are discussed here.

For those who want to explore the subject further, I have included a list of selected books and articles as further reading.

Finally, I have kept explanations as simple as I can; knowledge management involves many interrelated issues and some quite subtle ideas, but it need not be difficult to understand. Professionals such as architects, consulting engineers and doctors have always done many of the things it involves, and the challenge is not to do something fundamentally new, but to do it more consciously, with more understanding, and so more effectively. It is not rocket science. If one has enough of the basic ideas clear in one's mind, the implications emerge naturally: deciding what to do just takes effort to work things through, and the determination and patience to act on the answers. And it is worth the effort: as Drucker said, 'The basic economic resource . . . is and will be knowledge.'

None of my research would have been possible without funding from the UK Department of Trade and Industry. I am deeply indebted to all the organisations that have been my active partners in it over the years, and to my consultancy clients; without the lessons I have learned working with them to address the challenges of real-world knowledge management this book would not exist. I am especially grateful to Peter Oborn of Aedas, Chris Askew and Bill Gething of Feilden Clegg Bradley, Ashraf Michail of the BP/Bovis Global Alliance, Colin Rice of Edward Cullinan Architects, Andrew Cripps of Buro Happold, and Adrian Burton of Broadway Malyan. I cannot thank them and many others adequately for their openness about the sometimes untidy realities of practice, their willingness to take time out of pressured lives to debate ideas and try them out, their unfailing encouragement, and their kindness. Finally, I am eternally grateful to my wife Marion for casting a critical eye on drafts and for her understanding and tolerance. Nevertheless, despite all their invaluable contributions, any faults and shortcomings in this book are mine and mine alone.

Acknowledgements

I would like to thank Aedas, Broadway Malyan, Buro Happold, Edward Cullinan Architects, Feilden Clegg Bradley, Penoyre & Prasad and Whitbybird for allowing me to use the screenshots, photographs and diagrams that appear in their case studies.

I am also grateful to Arup and Colin Rice for allowing me to include the personal accounts of knowledge management initiatives which appear in the Arup and Edward Cullinan Architects case studies respectively.

Part One
Foundations

Chapter One
Introduction

A *New Yorker* cartoon of a few years ago shows an elderly man being introduced to a group of young staff, all looking up from their laptops: 'For those of you who don't know Mr Ingham – he's our institutional memory.' It is a neat encapsulation of a predicament that many organisations face today: the loss of accumulated expertise and dilution of corporate ethos as baby boomers retire and greenhorns flood in. And it is a reminder of deeper truths, too: that knowledge lives in people's brains, not in computers, and much of it can only be shared face to face, if at all. As the leading management thinker of the 20th century, Peter Drucker, put it: 'Knowledge is between two ears, and only between two ears.'

The loss of organisational memory and capability when long-serving staff leave is just one consequence of the general difficulty of sharing knowledge, and particularly the practical know-how that accumulates with experience. When knowledge is locked up in individual brains and local teams, unshared – as most of it is, in most organisations – wheels are reinvented, old mistakes are repeated, misunderstandings create new ones, and good practice stubbornly fails to spread. In professional services, practices that fail to pool their knowledge find economies of scale elusive, and that growth brings less competitive advantage than it should. Often, it seems to do little more than create a federation of small practices that share overheads.[1]

The scale of the waste from reinvention alone can be surprising. When I polled the staff of a large and highly successful architectural practice in a recent knowledge audit, only 25% thought they spent less than 10% of their time reinventing wheels, and 37% thought they spent over 20%. The average guess was 18%: that is typical. People's estimates of the time they spend looking for information are usually similar. The total effect in wasted time, lower quality and lost profit is

---

[1] One of the reasons why boutiques continue to prosper alongside giants.

considerable, and there are other prices to pay in missed opportunities to increase quality, reduce risk, and improve in other ways.

Not long ago, managers just shrugged their shoulders at all this – if they thought about knowledge at all – but the effect on corporate performance is becoming harder to ignore in an increasingly demanding and competitive world. At the same time, the trend towards larger, more dispersed and more complex organisations, and higher labour mobility, is causing knowledge to fragment more than ever and making it harder to share it. The Mr Inghams might be able to pass on their knowledge to half a dozen people round a table, but not to hundreds or thousands, spread across a country or across the globe.

Growing awareness of the value of sharing knowledge is the main reason for the proliferation of corporate knowledge bases, skills directories, communities of practice and other tools and techniques designed to make it flow around more freely. But these are not enough. Brains are not just passive repositories of knowledge; they *create* it by absorbing new experience and reshaping and extending old knowledge to accommodate it. Every time we tackle a new challenge, whether it is a hard-fought game of chess, a tricky design problem, or a meeting with a difficult client, we have an opportunity to learn how to do better on the next occasion we meet a similar situation. That is what turns the theoretical knowledge we acquire at university into practical competence, and develops junior staff into respected seniors, the most able into experts, and run-of-the mill firms into industry leaders. But few people outside psychology faculties think consciously about the processes involved, and as a result most personal and organisational learning is subconscious, haphazard, and more or less inefficient. That is just as wasteful as poor knowledge-sharing. Research has shown convincingly that the top experts and the top firms are those who accumulate the most experience, and learn most effectively from it.

When firms compete in a free market it is the differences between them that make one succeed more than another; the common factors merely define the baseline for entry. That makes the knowledge that comes from experience particularly valuable: its uniqueness makes it a key differentiator, whereas other knowledge is available to everyone for the price of a journal subscription or a course fee. Toyota's competitors can hire engineers from the same universities, buy the same books, and even tour its factories, but they do not have access to its unique experience, and few have learned so much from their own, or shared what they have learned so effectively. It is the knowledge that Toyota has accumulated in its workforce's heads, in company documents, in patents and in other forms – what Thomas A. Stewart called 'intellectual capital' – that has made it the most profitable volume car maker in the world.

To prosper in the 21st century, organisations of all kinds will need to become much better both at creating new intellectual capital and at using what they already have. This will require two things: *understanding* of what knowledge is and how it flows around, and active *management* of the processes of learning, sharing and the accumulation of corporate knowledge – in other words, knowledge management.

Paradoxical professionals

You would expect professional services organisations to be among the first to embrace knowledge management (or 'KM'); after all, knowledge is their stock in trade, and their staff and what they know are their largest asset. But no: with the notable exception of management consultancies (who *were* among the first), many of them have barely started. Who were the early adopters? The US Army, Toyota, Ford, Canon, Siemens, Chevron, BP . . . all organisations with huge assets of other kinds. To understand the paradox we need to look at how knowledge management has developed. That is worth doing because it shows why the time is ripe for professional services such as architecture, engineering, surveying and medicine to follow their lead, and why simply copying what they do will not work. It turns out that there are good reasons why the early adopters were fertile ground when professional services were not. Fortunately, it can be smart to be a late adopter: knowledge management is harder than it looks, and it helps to be able to learn from other people's successes and failures.

Knowledge management has had a meteoric rise. Before about 1995 the term was almost unknown, though some of the central ideas were already around under names such as 'organisational learning' and 'the learning organisation'. Today, it is familiar in the boardrooms of all kinds of organisation, across the world. The number of academic papers on the subject quadrupled between 1995 and 1997 and again by 1999, and books and articles on both theory and practice began to proliferate at the same time. Nonaka and Takeuchi's seminal *The Knowledge-Creating Company: How Japanese companies create the dynamics of innovation* reached the bookshops in 1995, and Stewart's *Intellectual Capital: The new wealth of organisations* appeared a couple of years later. Today a search for 'knowledge management' in Amazon.com books produces over 9000 hits; a general search on Google produces over 9 million.

Numerous organisations have taken up the idea and reported successes, often crediting it with major improvements in productivity and capability. As early as 1997 the Chief Executive of BP, John Browne, told a *Harvard Business Review* interviewer that improving learning and knowledge-sharing had generated $4 *billion* worth of permanent improvements in his company over the previous five years. When the Economist Intelligence Unit surveyed senior executives worldwide in 2005 and asked 'Which of the following areas of activity offer the greatest potential for productivity gains over the next 15 years?', knowledge management was the most popular choice by a wide margin. Assessing the changes likely in the global economy, industry and corporate structures over the same period, the EIU identified KM as one of the five principal trends, and concluded that improving the productivity of knowledge workers through technology, training and organisational change would be the major boardroom challenge of the next 15 years.

But there is another side to this rosy picture of progress, success and promise. Bain & Company have polled business executives almost every year since 1993 to see how widely various management tools are used, and how satisfied people are with them. By 2006 nearly 70% of the organisations surveyed reported using KM, with more planning to start in 2007, but only 17% reported being 'extremely satisfied' with it, and 16% were 'dissatisfied'. It ranked in the bottom 20% of tools for average satisfaction – as it has done every year since it was first included in Bain's survey. Satisfaction, of course, is a measure of the gap between expectation and achievement, and low satisfaction might only reflect unrealistically high expectations. That probably is a factor, but other evidence – and my own experience – suggests that low achievement *certainly* is. Booz Allen Hamilton estimated some years ago that only one KM programme in six achieves 'very significant' business impact in its first two years, half achieve 'small but important' benefits, and the remaining third are essentially failures. I suspect that little has changed since, despite the flood of advice in papers, books and conferences.

So is knowledge management a runaway success and a strategic priority for late adopters such as professional services, or is it a classic case of the emperor's new clothes – a deception nobody dares expose? I think it is something of both: a strategic priority and a success when realistic expectations and effective implementation coincide, but a disappointment when they do not. And it is too often made out to offer more than it really can, and to be easier to implement than it really is. To understand what it has to offer professional services we need to look beyond the generalisations of international, cross-industry surveys and consider what people mean when they talk about 'knowledge management', and why they continue to have such high hopes for it.

KM is a highly elastic concept, and it means very different things to different people. Software companies sell shrink-wrapped applications as 'knowledge management solutions' (none of them are!), and consultants and academics have described it in terms such as 'making the best use of the knowledge the organisation has got', 'the capacity to take effective action', 'about how to get people to work smarter', and even 'not the management of knowledge'. In practice, 'knowledge management systems' often turn out to be little more than old information management systems rebranded with a fashionable name, or a collection of procedures and IT tools that hardly anyone uses. With such a wide variety of usages, making sweeping judgements about it is like making judgements about transport without distinguishing between cars, boats and planes, or between what is being carried, and where. Further, the fuzziness of the concept makes it difficult for managers to form a clear vision of it, what it entails, or what to expect of it, let alone implement it successfully. And it is hardly surprising that many initiatives fall short of high aspirations such as 'making the best use of the knowledge the organisation has got'. We shall consider later what knowledge management can usefully mean in professional practice.

Despite its ambiguity, it is not hard to see why the idea of KM took off when it did and in the industries where it did, and why people still have such high hopes for it despite its mixed success in practice. Several key factors coincided for the first time in the 1980s and 1990s, and together they made the importance of knowledge in business clearer than ever before, and provided both the inspiration and the tools to do something about it:

- *Intangible assets* such as knowledge, patents and brands became the largest components of corporate value. The Brookings Institute has estimated that, as recently as 1982, over 60% of the market capitalisation of companies in the S&P 500 index was based on tangible assets such as factories, machinery and stocks; by 1992 the proportion had fallen to under 40%, and by 2002 it was less than 15%. The balance – today over 90% of value – is based on intangibles. The rise of companies such as Microsoft made the trend obvious to everyone during the 1990s, and acute business leaders were not slow to recognise its implications.
- *Globalisation* put pricing pressure on manufacturers, and at the same time showed the West how much Japanese companies were benefiting from their close attention to knowledge. By the 1990s Japanese industry had become a force in a range of major industries, making well-designed products with a production efficiency and quality that Western competitors struggled (and mostly failed) to match. Manufacturers used to dismissing Japanese goods as derivative and cheap-and-cheerful found that customers increasingly saw brands like Sony, Canon and Honda as premium

options, worth premium prices. Not surprisingly, Nonaka and Takeuchi found many eager readers when *The Knowledge-Creating Company* showed how much of their success was based on a culture of continuous learning and widespread knowledge-sharing. Combining Japanese production methods with cheaper labour in Korea, and later elsewhere, turned the competitive screw even further, and made it imperative for Western companies to adopt similar techniques.

- *Quality* became an imperative, too. I do not know why customers lost patience with faulty products, but they did. Perhaps it was just that the Japanese had proved that high quality was possible, or maybe it was a reaction to changes in manufacturing methods that made repair disproportionately expensive. Governments took advantage of the new possibilities to tighten regulatory standards for food quality, hygiene, waste disposal, energy efficiency, health and safety, and various other aspects of operations and products. Where regulation was not feasible, they cajoled. To improve standards in the construction industry, for example, the UK government sponsored a report on *Rethinking Construction* that lambasted it for endemic cost escalation, time overruns and defects, and called for 'radical improvement' in quality and efficiency. Together, higher customer expectations and tighter regulation made faults and mistakes matter more than ever before.

- *Growing size and geographic dispersion* meant that informal, intuitive communication ceased to work in many companies. As the Chief Knowledge Officer of Ernst & Young is said to have remarked: 'In the old days we used to yell down the hall "Has anyone done this before?", but you can't yell down a hallway of 75 000 people.' – especially when they are spread across a continent, or even a city.

- *Management styles changed.* The trend away from command and control styles of management towards flatter structures required knowledge as well as authority to be shared more widely.

- *Publications* such as *The Knowledge-Creating Company* and HBR's interview with John Browne, 'Unleashing the power of learning', brought three crucial elements together for the first time in a style that business leaders could understand and apply: an intellectual foundation for thinking about corporate knowledge, persuasive evidence of the impact that learning and knowledge-sharing could have on business performance, and practical tools for making it happen.

- *Personal computers became universal* for professionals, and the Internet established *universal standards for data exchange*. Together, these provided the technical means for people to communicate at a distance more freely than ever before, and for vast quantities of information to be stored and retrieved quickly and easily, from anywhere, by multiple users simultaneously.

These origins go a long way towards explaining why large, mostly manufacturing, corporations were the first to adopt knowledge management: they felt the pressures of changing business conditions first and most strongly, and they could relate to the early success stories. At the same time, they explain why professional services have lagged behind. They were sheltered from the greatest pressures (it is difficult to outsource the design of a school or treatment for a broken leg to China, and harder to compare competing architects than cars), and other pressures, such as rising customer expectations and challenging regulation and performance targets, were weaker and generally later to arrive in their markets. In future the differences look like becoming much less.

A McKinsey survey in 2006 asked respondents what single factor contributed most to increasing competitive pressure on their industry. 'Improved capabilities of competitors' – in other words, better knowledge or better talent – came top, chosen by 25%, followed by 'more low-cost competitors' (23%). Ten per cent chose 'growing size of competitors', 8% 'regulatory changes' and 5% 'rising consumer awareness and activism'. These are all as recognisable in contexts such as construction and medicine as in other industries: faster learning and making better use of existing knowledge are rapidly becoming universal imperatives. The Economist Intelligence Unit was surely right to conclude that knowledge management will be one of the principal trends in affecting business through to 2020 – and nowhere more so than in professional services.

New context, new issues

Even though they are increasingly subject to similar competitive pressures, professional services still differ from manufacturing companies in many ways, and will continue to do so. Expectations of KM and the way it is approached need to differ too. One of the key lessons from the past 10–15 years is that although the underlying principles of organisational learning and of knowledge-sharing apply everywhere, and many of the same basic tools and techniques can be used, the details of their implementation need to be tailored sensitively to the organisational context in order to succeed. Mies van der Rohe's famous dictum that 'God is in the details' is just as apt for knowledge management as it is for his minimalist architecture. We shall consider the implications of this in later chapters, but it is worth pausing to review a few of the characteristic differences between professional services organisations and other industries, and their consequences. Unique rather than mass-replicated products, managers who also own the business and earn fees, project working and an ethos of individual autonomy all have implications for knowledge management, and on the whole they tend to make it more difficult to implement. These differences are a further reason for its late adoption in most professional services, and they need to be confronted to make it succeed.

Most industrial and commercial organisations develop products and then replicate them essentially identically and in large numbers – cars, TVs, PCs, socks, steel bars, barrels of oil, tonnes of aggregate, insurance policies, retail transactions, train journeys, you name it – whereas professional services organisations typically deal in one-offs such as buildings, medical treatments, and consultancy projects. This difference has several consequences.

The most significant is that volume replication multiplies the value of improvements, particularly in operational efficiency and product quality, and creates the possibility of big wins. Even one new idea, or the transfer of a good idea from one factory, office or shop to others, might repay the annual cost of a company's KM programme.

The scale of potential benefits can easily justify substantial investment in seeking improvements to individual products and processes. A structured programme of learning and knowledge-sharing at BP focused on oil refinery refurbishments, for example, cut direct costs by 20%, reduced the time they took by 9 days, and produced a longer-lasting result – a total saving of nearly $10 million in each refurbishment, potentially repeated every 4–5 years and multiplied by around 20 refineries worldwide. Wins like this make both a strong business case for KM and good stories that can be a great help in convincing the indifferent and the sceptical. It is more difficult to justify generous investment in KM, and to motivate staff to make it work, in professional services, where the benefits are typically indirect, diffuse and largely unquantifiable, and big wins are almost impossible.

The role overlap between ownership, management and revenue-earning that is common in professional services is another factor that tends to make progress with knowledge management more difficult than it is in industries where they are separate. Its effects are particularly evident in medium-sized firms where ownership is shared relatively evenly between a dozen or more working partners or directors. Overlapping ownership and management puts decisions in the hands of people whose personal income is much more directly affected by short-term profitability than it is in quoted companies, and it may make investments in company-wide initiatives dependent on consensus between a dozen or more people. That is bad news for activities like knowledge management, which offer benefits, however considerable, that are hard to pin down and may take years to realise, in return for immediate costs, however small.

Two of the central tenets of behavioural economics (which won Daniel Kahneman a Nobel prize in 2002) are that most people are loss-averse – they will forgo the possibility of substantial gains in order to avoid losses, and put more effort into avoiding a loss than into securing a gain – and that they put undue weight on near-term events and too little on far-off ones in making decisions. Even after an initial decision has been made to invest in KM, role overlaps can be a continuing obstacle to progress. The principal cost of knowledge management is in staff time, and even when intentions are good it can be hard for people at all levels – and particularly management – to wrench themselves away from more enjoyable, revenue-earning activities. This is an instance of a widespread management problem that Stanford professors Jeffrey Pfeffer and Robert Sutton christened the 'knowing–doing gap', and we shall return to it later. Further, when firms operate more like a collection of independent baronies than a unified organisation, as is not uncommon in professional services, a local equity-sharing director unconvinced by knowledge management can completely block progress on his patch. In an environment like this, even appointing a dedicated knowledge manager is unlikely to make much difference. In a discretionary, non-fee-earning activity and without either professional standing or equity his position is too weak.

Dealing in one-offs almost inevitably necessitates project working, another characteristic that distinguishes most professional services from other industries: design the building, complete the assignment, treat the patient, and move on to the next. The cessation of revenue from each project when it finishes, the variation between them, and the creative professional's inner drive to try something new even when repetition might be more economic, all lead to a disinclination to look back systematically at completed projects in order to learn from them, let alone to make any effort to share lessons learned. Looking back costs money, a sacrifice of personal time, or both, and the lessons may be irrelevant in the next project. This is completely different from a typical manufacturing situation, where there is a conscious effort to make each new product an improvement on its predecessor, and to cut the cost of producing it, by identifying product weaknesses and process inefficiencies, finding ways to eliminate them, and mining competitors' products for good ideas.

All these obstacles can be overcome by leaders and managers prepared to make difficult decisions: accept the possibility of a small short-term reduction in income; make any necessary financial investments; delegate in order to clear personal time for knowledge management; give staff budgets for KM activities; make activities such as project reviews happen. But there are other obstacles that are less amenable to managerial determination. Professionals such as architects, consulting engineers and doctors are educated to expect considerable autonomy, and they are apt to believe that six or more years studying their discipline in university and in post-degree training has provided all the knowledge and skills they need. They are often reluctant to believe that anyone else can know better than they do, and strongly resistant to anything they see as interfering with their professional independence or creative freedom. Few professions have any tradition of looking elsewhere for ideas when people believe their existing knowledge is adequate.

The consequence is that many professionals search out information and advice only when they have to, and most tend to regard knowledge resources as a last rather than a first resort. Evidence-based medicine has only recently been accepted by doctors, and architects still show little inclination towards evidence-based design. Attitudes like these are far from unknown in other industries, but they are most deeply entrenched in the professions. A radical increase in learning and knowledge-sharing in an environment like this requires deep cultural change, and that poses a major challenge for business leaders who want their firms to use knowledge better.

Professional services organisations that have been late in adopting knowledge management, then, have not been perverse, but they would be perverse to delay much longer. As the management theorist Karl Sveiby has put it:

> Managers often have an unconscious and tacit mindset that is coloured by the values and the common sense of the industrial age. To see another world, they need to try to use a conscious mindset such as the knowledge perspective.

There is an overwhelming case for making KM a strategic priority: in the short to medium term to improve competitiveness, and in the longer term as a prerequisite for survival. There is much that can be learned from the way in which other industries have taken it up over the past 10–15 years, but professional services differ from them in ways that make blindly copying their approaches, tools and techniques unlikely to succeed; they need to be adapted to suit the different environment. And knowledge initiatives will stand or fall largely on the clear thinking and determination of leaders and managers.

What is in this book

This book has been written principally for partners, directors and managers (all 'managers' from now on, unless the distinctions are important) in architectural, engineering, surveying and property consultancies who recognise the importance of organisational learning and knowledge-sharing for their future success, and want their firms to be better at both. Despite the focus on construction, I hope the issues it discusses will strike chords, and the ideas it presents will be helpful, for managers in other professional services as well, in both the public and private sectors.

It is intended equally for readers who have got no further than putting knowledge management on the 'to do' list, for those who are struggling to create a KM strategy or to make a knowledge initiative work, for those who want to overhaul existing tools and processes that no longer seem fit for purpose, and for those who want to improve further processes that already work well. Fundamentally, of course, these positions are all the same. Learning and knowledge-sharing are as old as the human race, and every organisation today has informal working practices, formal procedures and IT systems designed to assist them in one way or another. Only entirely new organisations have the luxury of starting with a clean sheet.

This book does *not* address the handling of operational documents such as correspondence, contracts, schedules, specs or drawings, or business information such as time sheets, personnel records and accounts. These contribute only indirectly to knowledge, and the specialised software that is designed to store them, make them readily accessible, enforce version control and so on (excellent as it may be for its purpose) has little relevance to the management of knowledge and the creation and use of intellectual capital.

This first part, *Foundations*, goes on in Chapter 2 to set the scene by reviewing knowledge, learning, knowledge management, and what they mean in a professional services context. Chapter 3 discusses how the aspirations and operational focus of an organisation define priorities for learning and knowledge-sharing. Chapter 4 addresses an issue that many books on knowledge management ignore, but which seems to me to be among the most crucial: why knowledge initiatives so often disappoint or fail entirely. Chapter 5 discusses the crucial importance of leadership in achieving success, and the other roles that need to be filled. Finally, Chapter 6 turns to practical details and explains how to use a knowledge audit to establish the status quo and set objectives for a knowledge initiative (whether aimed at radical change or minor improvement), and how to use the results to develop an action plan.

Part Two, *Tools and Techniques*, discusses the processes and IT tools that are most likely to be useful in professional services organisations. Chapters 7–14 deal respectively with workspace design, social networking software to help people with questions find people with answers, mentoring, processes for learning at the start and end of projects ('foresight' and 'hindsight'), communities of practice (CoPs), the role of written knowledge and the software tools associated with it, personal knowledge management, and the relationships and synergies between them all. These chapters draw on experience accumulated over the past 15 years or so with the various tools and techniques in many kinds of organisation across the world, and discuss their strengths and weaknesses and how they can be tailored to suit the particular needs of professional services. Several chapters go into specific practical detail, but they are not recipes to be followed slavishly; rather, the detail is included to help readers visualise more clearly what the various tools and processes entail, and to provide a starting point for thinking creatively about them.

Part Three, *Knowledge Management in Practice*, describes some of the things that over a dozen of the most successful and managerially innovative companies in construction have done to improve their learning and knowledge-sharing. These are based on two research projects carried out between 2001 and 2005 in which I had the privilege of working closely with and advising them as they variously developed knowledge management strategies and implemented and tested new processes and tools. Most of the firms involved are professional practices, either architects or consulting engineers, but they also include the UK's largest airport operator (BAA), the BP/Bovis Global Alliance, a leading housing association, and others. They all started from different positions, and they followed a remarkable variety of paths. I am grateful to all the firms represented for their willingness to let me accompany them on their journeys, learn with them, and publish the details of what they did (and do) so that others can learn too from their difficulties and successes.

The *Epilogue* speculates on how organisational learning and knowledge-sharing might develop in the future.

Chapter Two
Knowledge at Work

We all think we know what knowledge is. It is such a pervasive part of life, and we say 'I know' so often without anyone asking for further explanation, that we rarely pause to consider what we really mean. But trying to manage an organisation's knowledge with only common usage as a guide is like trying to manage its finances with only a hazy idea of what money is – a recipe for disappointment, albeit with less immediately painful consequences. To be successful, knowledge management needs to be informed by a clearer understanding of the nature of knowledge, how it is created, and how people and organisations learn. Without that, managers are faced with a cascade of seemingly unanswerable questions when they try to choose tools and processes to match an organisation's particular needs, to get the details of their design and implementation right, or to take knowledge appropriately into account in other aspects of management. Why can't we just get all our experts to write down what they know? Isn't knowledge just information? What's special about face-to-face communication? What makes an expert? How can we make our new joiners productive more quickly? Why is it so difficult to transfer good practice from our London office to Newcastle? Why should we invite so many people to project reviews – the project leader can do it, can't he? Why should co-locating the design team help reduce project overruns? Why isn't it a waste of time for people to chat round the coffee machine?

How we learn

Research has shown that human infants develop a remarkably sophisticated understanding of causality, mechanics and other people's minds years before they acquire the language to talk about them. They quickly learn, for example, that for one brick to push another it has to be in contact with it, and they start to use objects as tools to extend their reach. They are surprised when an experimenter makes something happen that appears to violate causality. These are almost uniquely human abilities; even chimps fail to learn that a stick can be used to pull things towards them. We acquire fundamental knowledge like this by interacting with the world, observing it and thinking about what we have seen – by learning from experience – and that continues to be one of our most valuable sources of knowledge throughout life. It is, of course, the basis of all science and technology, too.

Causality shapes the way we understand the world, and the way we structure much of our knowledge. We start early to ask 'why' questions (as parents know to their cost), and the urge to find causal explanations continues to be strong throughout life. When observation fails to provide them we often invent them, and we may even invent entirely false memories to support them – all entirely unconsciously, of course. Memory is anything but the mental video recording we tend to imagine it to be: research has shown that we re-create our past every time we recall it, often slightly differently. And our behaviour is not controlled as much by the conscious, rational part of our brain as we usually assume. Our unconscious determines much more of our behaviour than most of us like to believe – it does well over 90% of our thinking, according to recent research. Sometimes our unconscious reinforces our conscious mind and sometimes it overrides it. It gives us intuitions that, research shows, are often remarkably accurate, gives experts abilities that they are unable to explain, and makes most of us avoid doing things we know we should. The American psychologist Jonathan Haidt has a nice metaphor: he sees the conscious mind as a rider on an elephant, a powerful and wilful beast that often decides to disobey the puny being who is trying to steer it. Managers need to understand something of both the rider and the elephant in order to create conditions, tools and processes that lead organisations to learn more from what they do, share individual knowledge more widely, and be more creative.

Learning is a remarkably varied activity: learning a PIN number, a poem, to recognise someone, about the American Civil War, to understand the equations of electrodynamics, to drive a car, to design a building and to manage a project are very different experiences. And whereas most of us can learn to recognise a face without conscious effort, remembering even four random digits requires some conscious attention, learning to drive typically takes 30–40 hours of practice, and it takes over 10 years to qualify as a neurosurgeon (in the UK). Despite this variety, the same memory processes are involved in all of them – and they are central to *using* knowledge, too.

All learning is based ultimately on sensory inputs from vision, hearing, touch, smell, taste and proprioception (awareness of the position and movements of one's own body). These are stored first in sensory memory, which is quite capacious but very brief – less than a second for visual memory and only a few seconds for sound memory. Sights and sounds (the only two inputs that concern us here) are processed locally to some extent, and then an interpretation of the parts to which we are paying attention is passed on to working memory. The hearing system, for example, has to disentangle a single stream of incoming sound into the voice on the phone, background chatter in the office, traffic outside and so on. The qualifications 'interpretation of' and 'to which we are paying attention' are important; we remember only a fraction of what we see and hear, and our memory of an event may be quite different from other people's, and even quite wrong.

Differences and errors in interpretation are the basis of sensory illusions such as the drawing that sometimes looks like a vase and sometimes like two profile heads facing each other, the Escher staircase that keeps on climbing as we follow it round, only to end up where it started, and the musical tone that goes on rising for ever. Failures of attention are famously illustrated by Harvard professor Dan Simons' 'Gorillas in our midst' experiment, in which about half of the people watching a short video of students playing with a basketball failed to notice a woman in a gorilla suit walking across the scene – even in a variant in which the gorilla stopped half way, turned towards the camera, and beat a tattoo on its chest. Everybody, of course, sees it when it is pointed out.

The visual system constructs its interpretations from numerous 'snapshots' in which the eye rests briefly on one point at a time before moving abruptly on to another, and studies of eye movement show that even these depend on what we are paying attention to. People asked variously just to look at a picture, to estimate the economic status of the people in it, or to judge their ages showed quite different patterns of eye movement as they subconsciously searched for different kinds of evidence. It seems that even at the most basic level – while data is still largely sensory, and before significant meaning has been attached to it – the way we perceive the world depends on what makes sense to us, what we expect to see (and hear), and what we choose to pay attention to, almost as much as on what is actually in front of us. No wonder early explorers brought back weird and wonderful accounts of what they had seen, learner drivers find busy towns so confusing, and witnesses can give such different accounts of crimes.

Effects like these are compounded by the characteristics of working memory, with which sensory memory works closely. This is both the next staging post for most of the things we see and hear on their way to long-term memory (some appear to have a more direct route) and the place to which we have to recall existing memories in order to use them in conscious thought. It is where we take the crucial step of attaching meaning to sensory inputs and encoding them into forms that can be stored in long-term memory, such as words. Information in working memory lasts longer than it does in sensory memory, but not much: about 10–15 seconds. To hold it there longer we have to refresh the memory by repeating it to ourselves. Without rehearsal, the proportion of people who can accurately recall a short, meaningless string of letters slumps from around 90% immediately after seeing them to less than 10% after 15 seconds.

George Miller showed 50 years ago that few people can hold more than seven or eight random independent items (such as random consonants) in working memory, however hard they try, and subsequent research has confirmed that this is about the limit of its capacity.[1] That is extraordinarily small for a channel through which most of our knowledge, memories and conscious thinking have to pass. It would be completely inadequate if memories were like DVDs (just one of which contains nearly 5 *billion* independent items of data), but they aren't. To overcome the capacity limitations of working memory, the brain makes each of the seven or so items it can handle at a time do more work than a byte on a DVD by ruthlessly discarding what it considers irrelevant, encoding information with extraordinary efficiency, and reusing existing knowledge when it can.

Memory of sensory experiences, of course, is only one, limited, kind of knowledge. Neuroscientists distinguish between episodic, procedural and declarative memories, respectively memories of events, motor skills such as driving a car, and verbalisable facts – in common parlance, information – such as names and mathematical methods. These can all be important in professional life (some of the most valuable skills involve combinations of all three) but declarative memory is the most commonly important, and the one that we use most in conscious learning. Research suggests that the keys to this are recognising patterns in what we see, interpreting it in the light of prior knowledge, and connecting it to prior knowledge. A random number generator asked for nine digits is no more unlikely to produce 111222333 than 736129554: both sequences are equally meaningless, but the pattern in the first one – short, simple groups of ones, twos and threes, arranged in a familiar order – makes it much easier to remember. Familiarity can be just as helpful. Many of us use numbers with a personal significance as PINs because they are easier to remember, and it would be much more difficult to learn how to use new software if we did not already understand menus, toolbars, scroll bars and common commands like 'open' and 'save'.

---

[1] And we can only change its contents about 18 times a second, so our conscious mind can only handle in the order of $7 \times 18 = 126$ bytes per second of data, a small fraction of the capacity of the slowest modem.

Concepts, language, mathematics, physical laws and rules of thumb are all aspects of these processes at work. They help us to see patterns in observations that would otherwise appear mysterious, to make sense of new experience and knowledge, and to share our understanding. In a famous early experiment on learning a century ago, two groups of children were asked to throw darts at a target underwater. Those in one group were taught about refraction beforehand, so they knew that the actual position of the target would be offset from its apparent position, and that the offset would increase with the depth of water; those in the other group were not. Both groups performed equally at first, gradually learning to adjust their aim to compensate for the water. But when the depth was changed the children who understood refraction were able to adjust their aim more quickly than the others, because they had a mental model of how the water was affecting what they saw. Both as individuals and as a society we build our knowledge and capabilities brick on brick, new on old, using simple ideas to create more complex ones.

Unfortunately, though, understanding an explanation or spotting a meaningful pattern in our own experience is no guarantee that we will remember it in the long term, or even tomorrow. Forming secure long-term memories usually takes practice: we need to recall our new understanding repeatedly over a period of time to develop the necessary neural connections. The more we recall it and link it to other experience and knowledge – by thinking about it, using it, or discussing it with other people, for example – the more securely we remember it, and the more easily it comes to mind when we need it. As Confucius is alleged to have said, 'I hear and I forget. I see and I remember. I do and I understand.' The doing is crucial.

What makes an expert

As people learn more about a field, and gain more experience in it, they become increasingly competent and, eventually, expert. Experts possess vastly more information, examples and mental models than novices, more richly interlinked, and this enables them to *think* differently, too. The difference is perhaps easiest to see in chess players. A novice looks at all the pieces he could move and considers consciously where they could go, what advantage each move might give him, and how his opponent might respond. If he has time and a good enough memory he might go on to consider his options for his next move after that, or even one move further again. Until the game is nearing checkmate, with only a few active pieces left on the board, this approach involves daunting numbers of possible moves and responses, numbers that escalate massively with each additional move considered. The mind boggles at the grandmaster Capablanca, playing a group of opponents simultaneously, taking two or three seconds to make his move (while everyone else has as long as it takes him to complete a circuit to make theirs), and winning every game. How could he possibly analyse so many possible moves, so quickly and so effectively? The answer is that he didn't. When asked, he is said to have replied: 'I see only one move ahead, but it is always the correct one.' If he was right, chess masters must think *very* differently from novices. Research suggests that (allowing for a little exaggeration) he *was* right, and that masters in every field really do think differently. But *how* do they think differently? What make an expert expert?

There is nothing fundamentally wrong with the amateur approach of comprehensively analysing possibilities except the limitations of the human brain, at least in a situation governed by a few formal rules, like chess. Computer chess programs that work in more or less that way can be quite strong players: fast processing and perfect memory make it possible for them to look much further forward than humans, and though their play lacks flair, it is effective. But tests show that grandmasters do not think faster, or have better memory in general, than other people, and yet they can still beat computers most of the time. Research has shown that they often do analyse possible moves, but no more of them than moderately competent players. However, they concentrate their analysis on the most promising ones – and sometimes (in Capablanca's case, usually) they can see the best one immediately.[2]

Grandmasters' years of thoughtful practice and study give them a hugely greater repertoire of remembered positions, moves and strategies than the average player, and that enables them to short-circuit analysis by recognising patterns and recalling ready-made solutions. They make extensive use of their long-term memory while weaker players have to make do with working memory. Experts in other fields do the same: when presented with a task, their experience and deep understanding enable them to see the features that matter, ignore irrelevancies on which the less expert tend to get hung up, home in on the factors that are likely to lead to a solution, and make progress quickly. They have the kind of intuition that Nobel prize-winning psychologist Herbert Simon called 'analyses frozen into habit and into the capacity for rapid response'.

Recent research on mice brains by Joe Tsien and his colleagues at the University of Boston suggests that this ability is inherent in the way that their – and our – brains form memories. Using sophisticated experimental techniques, instrumentation and mathematical analyses

---

[2] Quite a contrast to IBM's Deep Blue, which needed to evaluate 200 million moves a second to become the first chess-playing computer to beat a reigning world champion in 1997.

to study, for the first time, the behaviour of large *groups* of individual neurons, Tsien's team have shown that the brain encodes the key features rather than specific details of experiences, and in so doing extracts from them 'general information that can be applied to a future situation that may share some essential features but vary in physical detail'. This is the basic building block of expertise, and it is not difficult to imagine that expert knowledge is essentially an accumulation and combination of memories like these from prolonged and concentrated practice and study.

In essence, then, the important differences between novices and experts are that the experts:

- Possess vastly more knowledge of all kinds, including examples, general laws, manipulative tools, physical skills, and specialised language, as a result of years of experience and thinking about their field, and this is more richly interlinked and understood in terms of higher-level concepts
- Notice patterns in situations and problems that novices miss, ignore irrelevancies, focus on key issues, have generally accurate intuitions, and deploy the most relevant concepts, parallels and tools
- Make extensive use of long-term memory, recalling key knowledge automatically and without conscious effort.

Experts think both more effectively and more efficiently than novices, so they can achieve better results with less effort, and successfully tackle tasks that would stump novices completely. And because they know more, they learn faster and remember better, too. Knowledge breeds knowledge.

Expert knowledge is the reward of years of concentrated effort. Research suggests that it typically takes around 10 years to acquire, whatever the field. It is the combination of a vast mental stock of information and a rewired brain, and there is no short cut to acquiring that. However, the process can be accelerated by combining teaching of organising principles with guided and judiciously graded experience in a succession of situations each near the limits of the learner's current expertise, or just beyond. Simply accumulating experience without continual stretch does little to increase expertise, as the static skills of millions of weekend sports players shows. It is no accident that teaching combined with continually stretching experience is essentially the pattern that has evolved for the education of doctors, lawyers and architects. It is rarely continued into practice in any organised way, but there is no reason in principle why it should not be. And research shows that application matters much more than innate talent: given reasonable intelligence and sustained effort, it appears that anyone can become an expert in almost any field.

Understanding how we learn and what makes an expert provides a basis for professional services organisations to make graduate entrants productive more quickly, enable all their staff to develop their expertise faster and farther, create conditions for stars to shine, and generally build up intellectual capital. Those are worthwhile and realistic ambitions, and knowledge management provides the strategic framework, the tools and the techniques for achieving them.

Varieties of knowledge

Knowledge we can and can't explain

One of the tantalising features of expertise is that, like Capablanca, experts are often unable to explain how they reach their conclusions; because so much of their thinking is subconscious they simply do not know. Indeed, it may well involve a kind of parallel processing that would be impossible to explain as a sequence of steps for someone else to follow. We all have areas of expertise, if only small ones, and as philosopher Michael Polanyi put it, we all 'know more than we can tell'. He called knowledge of this kind 'tacit' to distinguish it from the 'explicit' knowledge that can readily be explained, written about and taught – knowledge such as facts, and techniques for working with them like language and maths.

Polanyi wrote about 'tacit knowing' as a process rather than a form of knowledge, and emphasised the importance of factors such as beliefs, habits and culture, which are essential parts of our capability without our being conscious of them. In reality, of course, 'tacit' and 'explicit' are the ends of a spectrum with infinite gradations of awareness and communicability in between. Tacitness is a matter of degree: knowledge is tacit to the extent that it is towards the end of the spectrum that is personal, unconscious, stems from learning and experience, is rooted in specific contexts, and includes paradigms, viewpoints and beliefs as well as intellectual and performance skills. Most professional knowledge has elements like that, alongside others that are entirely explicit.

Formal education concentrates on equipping graduates with explicit knowledge. The knowledge they need to become effective professionals comes later from practice, and it is more tacit. It is many years' accumulation of this largely tacit, experiential knowledge that makes a senior partner more capable than a young trainee. An organisation's collective knowledge is like an iceberg, with the explicit knowledge above water dwarfed in volume and value by the tacit knowledge under the surface (Figure 2.1).

Nonaka and Takeuchi argued in *The Knowledge-Creating Company* that it is the spiral of interactions between the two that is the basis of corporate innovation: individuals learn and share their tacit

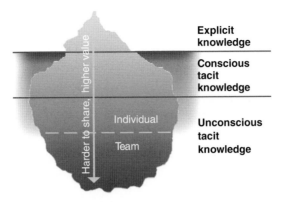

Explicit
knowledge

Conscious
tacit
knowledge

Unconscious
tacit
knowledge

Individual

Team

Harder to share, higher value

Figure 2.1 The knowledge iceberg.
The explicit knowledge that can be seen in an organisation's procedure manuals and knowledge bases is dwarfed by the tacit knowledge hidden in people's heads. Some of this is readily available to anyone who asks them a question, but there is even more that they could explain only with difficulty, or not at all. The highest levels of expertise are deeply tacit, unconscious, and invariably impossible to articulate. Knowledge possessed only collectively by teams can be equally valuable, and equally inaccessible. The competitive value of knowledge tends to increase with depth in the iceberg.

knowledge in face-to-face interaction ('socialisation'), it is made explicit ('externalisation'), further knowledge is created by combining explicit knowledge ('combination'), and people learn from this and make it part of their own expertise ('internalisation'). That is debatable, but there is no doubt about the value of tacit knowledge, both as a conceptual help in understanding organisations, and in reality as a major factor in organisational capability. Most of a professional practice's collective knowledge is tacit to some degree. This tacit knowledge is its greatest business asset; it is largely invisible; and it is as important for innovation and success tomorrow as it is for performance today.

The importance of tacit knowledge has both good and bad consequences for business. It means that the most valuable knowledge is hard to spread around an organisation, but it also protects it from becoming widely available to competitors. That makes tacit knowledge a key source of sustainable competitive advantage, and one of the main aims of knowledge management is to create more of it and make it flow around more freely.

Talk is magic

The most highly tacit knowledge cannot be shared at all; it can only be acquired from experience. At the other end of the spectrum, purely explicit knowledge can in principle be made widely and readily available by documenting it in (for example) a formal procedure, a technical note or a knowledge base, and it can be acquired by reading.

In between, there is a broad band of knowledge that is too tacit to communicate effectively in a textbook or a corporate knowledge base, but is communicable to varying degrees face to face, or (usually less well) by phone or email. This involves more than just a flow of words from one person to another – from a teacher to a learner – that could in principle be written down; it is an interactive process between them, which demands physical engagement and focuses attention in a way that documents cannot. In conversation, teachers can:

- Be *selective* and *pertinent*, focusing narrowly on learners' specific interests and sparing them the need to find and extract relevant information themselves – something they may not have the time or expertise to do
- *Interpret* information to suit the context in which the learner wants to use it
- Be *responsive*, providing information piece by piece as the learner absorbs it, at an appropriate pace
- *Adapt* material to suit the learner's existing knowledge
- *Listen* to learners' explanation of their understanding, and explain it another way if that is wrong
- *Demonstrate* physical actions
- Make freer use of *anecdotes and stories* to help understanding than would usually be appropriate in documentary sources
- *Guide experience and experiment* so that learners can try things out and make the words they have heard come to life.

The learner, meanwhile, can:

- Adjust or extend the original question ('And what would happen if . . . ?')
- Ensure that new understanding is correct by verbalising it ('So you mean that . . . ?') or by reproducing physical actions, and getting corrective feedback.

Echoing new knowledge in this way also helps learners to build mental connections with prior knowledge, and fix the new knowledge in their brains.

Recent research has revealed other unique attributes of learning face to face. It turns out, for example, that watching an action demonstrated activates the motor areas in our brain that we would use to perform it ourselves. In effect, we practise the action in our heads, giving us a much better starting point for remembering it and performing it successfully later than the ambiguities of written descriptions.

The overall advantages to the learner are considerable. Selectivity, pertinence and interpretation all save time searching and wandering down blind alleys; responsiveness, adaptation and listening help avoid misunderstanding; anecdotes, stories and active participation all make new knowledge more memorable. The benefits are not solely one way, either. Teachers benefit, because verbalising their own understanding and having aspects of it questioned helps to develop it further. And when, as often happens, two or more people each have incomplete knowledge, discussion can often reveal pieces that fit together, spark new mental connections, and lead to new understanding and new ideas.

Face-to-face interaction, then, is a uniquely powerful – and sometimes the only – way to share many kinds of knowledge, from the simplest to the most sophisticated. It is one of the best ways to stimulate new thinking and ideas, too. Most of us would have had difficulty learning how to tie a shoelace solely from pictures, or how or do arithmetic from a book. At the other end of the scale, psychologist Mihàly Csikszentmihàlyi found, while studying high achievers, that a high proportion of Nobel laureates were the students of previous laureates: they had access to the same literature as everyone else, but personal contact made a crucial difference to their creativity. Within organisations these qualities make conversation both a crucial vector for high-level expertise and the most important way of sharing mundane, everyday information. Few, though, recognise its importance, or do anything to exploit it more effectively; its very naturalness makes it invisible. It is another of the aims of knowledge management to bring it more into play.

But one-to-one knowledge-sharing does not scale easily. The flow is largely limited to people who know each other, often making a large organisation more like a collection of isolated knowledge villages than a vibrant knowledge city. And it is burdensome and inefficient for recognised experts to have to repeat explanations to numerous people. That makes face-to-face interaction the medium of choice for the kind of informal and local knowledge-sharing that occurs naturally in an office, for spreading high-value tacit knowledge that cannot be communicated in any other way, and for stimulating knowledge creation, but *not* for disseminating explicit knowledge widely and quickly throughout an organisation (unless it is very small). It is much easier to share knowledge that *can* be written down across a large and geographically dispersed practice when it *has* been written down.

. . . but the pen is mighty, too

Information technology makes it as easy to share recorded knowledge across large and fragmented organisations as across small ones, but the process of capturing it in words and pictures is a stumbling block. It only has to be recorded once, but that usually takes more time, effort and skill than simply explaining it in conversation. Writing it down can seem just as burdensome to authors as explaining it several times over. Raw material may have to be collected by studying other documentary sources, eliciting it from other experts, and perhaps research; sources may have to be combined and reconciled; general principles may have to be extracted from special cases; and the result always has to be turned into coherent, readable prose, clear illustrations, lucid commentary, or whatever the medium demands. As anyone who has written an article for a professional journal will know, even that is a non-trivial task for all but the most fluent, and producing a substantial manual or making an instructional video is much more so. As a result, relatively little knowledge is recorded in professional practices today. When explicit knowledge is documented it is most often with commercial publication in mind, or as a collaborative effort by bodies such as professional institutions and standards organisations.

Knowledge that is available in the public arena (including the knowledge available in formal education) offers relatively little competitive advantage, because anyone can buy it, usually quite cheaply. However, most organisations possess a considerable amount of potentially recordable knowledge that is unique, and unique knowledge is one of the principal sources of competitive advantage in professional services. There is a considerable opportunity cost in leaving it in people's heads, unrecorded, and that cost grows rapidly as organisations expand and fragment, as face-to-face sharing becomes increasingly inadequate, and as increasing numbers of people leave and join. Until recently this was unavoidable, but new software tools such as wikis are at last making it feasible for organisations to capture more of this kind of knowledge, at reasonable cost, in forms that make it accessible to everyone. This is yet another aim of knowledge management.

Just-in-case or just-in-time?

The distinction between tacit and explicit knowledge is not the only one that helps in understanding organisational knowledge. The difference between just-in-time and just-in-case knowledge is also important. In the course of everyday work people ask a colleague a question or look in a manual only when they need to fill a gap in their own knowledge. They want an answer that is relevant, concise, reliable, and available *now*: just-in-time knowledge. Education, training, and reading the latest professional journal, on the other hand, provide just-in-case knowledge: knowledge that has no immediate use, but (authors and readers hope) *may* be useful in the future. Just-in-case knowledge needs to be interesting and memorable, but speed of access and conciseness are unimportant. Just-in-time and just-in-case knowledge differ in other ways, too. Some of these are shown in Table 2.1.

Historically, most organisations have given conscious thought only to providing just-in-case knowledge, through formal training.

	Just-in-time	Just-in-case
Purpose	Problem-solving	Developing knowledge and skills
User requirements	Quick access, relevance, conciseness, clarity Does *not* need to be particularly interesting, memorable or polished	Interest, memorability, polished presentation Does *not* need to be particularly quickly accessible, relevant to work in hand, or concise
Location	In colleagues' heads or documents	Usually in documents
Form	Organised for reference (documents) Narrow topic focus Stand-alone, bite-sized pieces More often factual information than skills	Organised for learning Broader topics Longer units, narrative style Discusses principles and techniques as well as offering facts
Identified by	Question (person-to-person) Topic index, search engine, hyperlinks (recorded knowledge)	Title, contents list, summary

Table 2.1 Comparing just-in-time and just-in-case knowledge.

Knowledge management developed in part as a response to the dawning recognition that it is even more important for them to provide just-in-time knowledge. A rounded corporate knowledge strategy, of course, needs to consider both.

Putting the pieces together

The knowledge jigsaw

Clearly, knowledge is a more multifaceted, complex and slippery concept than most people imagine and most organisations have historically assumed. Indeed, philosophers are still arguing about what 'knowledge' really means. The *Oxford English Dictionary* offers seven principal modern usages, and several more that are specialised or obsolete. The only important similarity between knowing *that* William of Normandy invaded Britain in 1066, knowing *that* you are awake, knowing *how* to bake a cake, knowing *how* to beat Kasparov at chess and simply *knowing* your best friend is that, as Peter Drucker put it, they are all 'between two ears'. Creating the conditions and providing the tools to make the best – or even just better – use of knowledge in professional services organisations is a major challenge. Before considering how this can be achieved in practice, we need to pull together the ideas we have discussed in this chapter and see what they add up to – what 'knowledge' really means in the context of knowledge management.

Organisations exist to do things, and the only kinds of knowledge that matter are those that help in this. In organisations, therefore, 'knowledge' is first and foremost the ability to do something, a capability or competency – what knowledge management pioneer Peter Senge called 'capacity for action'. 'Knowing how' to do things – know-how – is also one of the most familiar kinds of knowledge in everyday life, though entirely different from the 'knowing that' which is the mainstay of general knowledge quizzes. Capacity for action calls for a combination of explicit and tacit knowledge, of information, beliefs, and intellectual, emotional, sensory and motor skills, conscious and subconscious. When we say we know how to write a project proposal, analyse risk, design a building, lead a team or manage a business we mean that we have the combination of the elements that we need in order to do those things. I find it helpful to think of know-how like this as a knowledge jigsaw (Figure 2.2).

The jigsaw model makes several important points about this kind of knowledge:

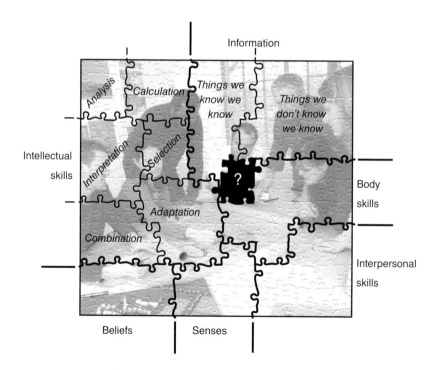

Figure 2.2 Know-how is like a jigsaw.

- It may take many, varied pieces to create a capacity for action.
- Different competencies call for different combinations of pieces.
- All the pieces have to be there to complete the picture and enable people to act effectively; the lack of even one will make them less effective, and may make action impossible.
- It may take only one, or a few, additional pieces to complete a picture, and enable people to do something new.

The jigsaw also suggests a way to resolve the paradox at the heart of knowledge management: that most management and knowledge theorists agree with Peter Drucker that 'knowledge is . . . only between two ears', and yet to share it we somehow have to represent it in one of the forms that can exist *outside* the human head, such as spoken words, a printed page or an electronic knowledge base.

These latter forms are usually called data or information, and theorists continue to argue about the relationships between data, information and knowledge, and how one leads to another. Most interpretations are variations on the proposition that isolated facts are only data; when they are put in a context that makes them meaningful they become information; and information becomes knowledge when people absorb it and it becomes part of their mental tool kit. Individual lines in a profit and loss account, then, are only data; together, they become information; and when the CEO reads them they become part of the knowledge that (hopefully) helps him or her to make good management decisions. This looks clear, but in practice it is hard to maintain a distinction between information and knowledge in these terms. In their classic *Working Knowledge*, for example, Davenport and Prusak suggest that

> Knowledge is a fluid mix of framed experience, values, contextual information, and expert insight that provides a framework for evaluating and incorporating new experiences and information. It originates and is applied in the minds of knowers.

and then go on to say that 'it often becomes embedded . . . in documents.' Even if the distinction was sustainable, its practical implications for knowledge management would be far from obvious.

The jigsaw model offers a different way to think about all this, and one that I find much more helpful. Giving someone information that completes a knowledge jigsaw in their mind gives them a new 'capacity for action' – new *knowledge*. In this case, therefore, sharing the information is equivalent to sharing the knowledge. For practical purposes, information that completes a jigsaw for someone *is* knowledge, and we can reasonably *call* it knowledge. Information that does *not* complete a jigsaw is only information.

Accepting that capacity for action and the information needed to complete a knowledge jigsaw can both be called 'knowledge' does not mean that knowledge and information are the same, but only that *some* information can be regarded as knowledge in *some* circumstances. As Stewart puts it in *Intellectual Capital: The new wealth of organisations*, 'One man's knowledge is another man's data.' Whether information is equivalent to knowledge depends on the users; it is knowledge only when it completes a picture for them, and gives them a capacity for action that they lacked before. Like beauty, knowledge is in the eye of the beholder.

I find that this really does help to clarify what sharing knowledge entails, and what knowledge managers need to do. It explains why one-to-one conversation and demonstration are such effective ways to share knowledge: as we saw earlier in this chapter, feedback enables the teachers to give the learners precisely what they need to complete their personal jigsaws. And it shows that, to share knowledge successfully through media such as technical papers and knowledge bases, knowledge managers, system designers and authors need to start by finding out what their readers do and don't already know.

Most just make assumptions, and the results are discouraging. One of the commonest reasons for disappointment with knowledge bases is that the content management has chosen to provide is not what users actually need – it is information that *doesn't* complete useful knowledge jigsaws. Similarly, as Gabriel Szulanski's research has shown, the main reason why so many attempts to transfer best practices have failed is that recipients were assumed to know things that in fact they did not – they were given information, but still left with gaps in their jigsaws.

In most organisational contexts, of course, the same elements recur in many knowledge jigsaws and are already part of many people's repertoire, so they often need only a few new pieces to give them new competencies. And even though people's individual knowledge and competency is lost when they leave an organisation, some of their capability can be preserved by recording critical information that only they possess, ready for others to use.

Knowledge management

How successful an organisation is at creating and using knowledge – 'the basic economic resource', according to Drucker – depends on how well the three fundamental processes we have been considering in this chapter are working: learning and the development of new knowledge and expertise, the sharing of tacit knowledge, and the recording of high-value explicit knowledge to create an organisational memory that is available to everyone. All three work to some extent without conscious attention. Most people gradually become better at their jobs; we all ask colleagues when we need the last piece for a knowledge jigsaw; and we dutifully file project papers, and from time to time find something useful in old files. But they can all be made to work much better with conscious, skilled management, and that is what we mean by 'knowledge management': it is what you do to make organisational learning happen, create a learning organisation, and build intellectual capital. If you want a definition, mine would be that it is *the conscious attempt by managers and individuals to make themselves and their organisations more capable by taking better advantage of opportunities to learn, and by sharing knowledge more effectively.*

That is a quite a packed sentence, and it is worth picking apart. The key concepts are:

- *Consciousness.* We all learn, share knowledge and organise information, but mostly without understanding or even thinking about the processes involved, and therefore (since they are quite complex activities) inevitably rather ineffectively. Understanding knowledge processes, and applying that understanding in managing both organisations and our own work, can make them work much better. Knowledge needs to be part of every manager's mindset.
- *Attempt.* There are all shades of grey between complete ignorance of knowledge management (the usual condition until a few years ago) and having a sophisticated and successful range of knowledge systems and practices – and between total failure and great success. There are no guaranteed recipes; all anyone can do is try, and keep on trying. The theory and practice of knowledge management look set to continue evolving and improving for many years.
- *Managers and individuals.* Knowledge management is something *people* do, not something done by an IT system – and it has to be done by people at all levels to work well. Management can enable and push, but success depends just as much on the development of good habits at the grass roots where most of the organisation's work is done. Everybody can become more capable at their own work by learning, sharing and using knowledge more effectively, and conscious understanding makes it easier to do so and more likely that this will happen. Management has the additional responsibilities of providing tools and processes to enable people to contribute to organisational learning as well as to their own, creating a supportive culture, and resolving

conflicts between knowledge activities and other priorities and pressures.

- *Themselves and their organisations.* Knowledge management needs to deliver both business and personal benefits to succeed: the business benefits to justify investment in creating and maintaining the process, and the personal benefits to motivate people to make it work.
- *Capable.* The basic purpose of knowledge management is to enable people and organisations to do familiar things better, and learn to do new things. It is that increased capability that produces the corporate and personal benefits: corporately, in better products and services, lower costs, lower risk, more satisfied clients, greater resilience, and ultimately sustainable profitability; personally, in work that is easier, more efficient and more successful, and in greater work satisfaction, and faster career advancement.
- *Learning* and *sharing.* All knowledge starts with learning, but the benefits are limited unless new knowledge is shared – ideally with everyone else in an organisation who might find it useful. Sharing, incidentally, depends equally on everyone both making their knowledge available to others *and* on using the knowledge made available to them, on both push and pull; it is not a one-way process.
- *Knowledge.* IT has revolutionised the possibilities for making information widely, easily and quickly accessible, and the obvious benefits of doing this have often made it a major (and often the only) aim of so-called 'knowledge management'. But the only point of sharing knowledge is to enable people to do things that they could not do before, and, as the jigsaw metaphor makes clear, this means that the information has to complete useful knowledge jigsaws for its users. Failure to see knowledge-sharing from the *users'* point of view leads to disappointment. Indiscriminate dissemination leads both to failure to provide the missing pieces people really need and to information overload.

Organisations that manage knowledge well develop superior prod-ucts and services, become better at innovation and production, have happier staff, please their customers more, grow, and become more profitable. Equally importantly, the range of their knowledge resources allows them to respond flexibly to circumstances, and makes them resilient in a changing world. The rest of this book discusses what good knowledge management involves in practice, and what some professional services firms have actually done to start making it a reality – to become truly learning organisations.

The best way to start any substantial knowledge initiative is to consider two fundamental questions: what are the organisation's aspirations and business strategy, and what do they say about the part that knowledge plays in its work? Understanding this gives clarity to all that follows: it shows how to shape knowledge strategy, tools and processes to make them serve business goals and enable investment in them to deliver the best return. The next chapter looks at how this can be achieved. We go on from there to consider why knowledge initiatives have so often failed in the past (Chapter 4), how the pitfalls can be avoided (Chapter 5), and (in the last chapter of Part One) how to assess the current state of an organisation's knowledge management, identify priority areas for improvement, and develop a plan of action.

Chapter Three
Strategic Frameworks

Knowledge is self-evidently a good thing, but that does not guarantee that investment in improving the way an organisation manages knowledge will pay off as handsomely as it has done for companies such as Toyota and BP. The visible costs may be small, but any substantial initiatives are likely to demand significant amounts of management time, from the chief executive down, and the business benefits have proved disappointing in many organisations. To deliver the best return, initiatives need to be guided by a clear vision of their strategic purpose, be shaped to meet specific needs, fit the organisational context, and be driven by real conviction and commitment.

Starting points

The kinds of knowledge that matter, where they come from, and how they are used vary widely. Scientific knowledge and the research literature are vital in the development of new drugs, and the testing and regulatory clearance needed to get them onto the market are both knowledge- and document-intensive. Neither matters to a supermarket chain, where efficient supply chains, logistics and shopping patterns come to the fore. Professional services organisations have more in common with each other than with drug companies and retailers, but there is still a great deal of difference in aspirations, structure, geography, management style and culture between an architectural cooperative of 40 (such as Edward Cullinan Architects) and an engineering plc with 4600 staff operating from 110 offices worldwide (such as WSP). This has profound implications for the objectives and practice of knowledge management; one size does not fit all.

There are obvious practical differences – Cullinan's staff can meet round one table, whereas most of WSP's staff will never see or speak to most of their colleagues, for example – and subtler ones too. Research at Portsmouth University has shown that architects and engineers typically have very different learning styles, with architects preferring to learn through direct personal experience, whereas engineers prefer abstract, broadly applicable principles and established rules. Culture can vary considerably, too. Some practices are more authoritarian and have more formal procedures than others, some are specialists while others are diverse, many have never given any thought to knowledge, and a lucky few (such as Arup) have had it in their corporate DNA since their foundation. Differences like these call for different approaches to learning and knowledge-sharing. The techniques discussed later in this book are those that seem to me to be most relevant in professional practices, but not all will be appropriate for every practice. And the step-by-step recipes for some of them are not intended to be prescriptive; they are simply intended to show how the theoretical principles translate into practicalities, to stimulate thought, and to provide a starting point for experiment and for developing systems that suit a practice's individual needs.

Knowledge management is not something that can be 'bought' or 'delivered'. It is a continuing journey with many hazards, and it needs to be directed by a compelling vision of knowledge's role in the organisation's work and future development. Experience in other industries suggests that its most successful exponents have several factors in common. They understand what kinds of knowledge matter in their business, and how using knowledge better, and learning faster, can support their business strategy; they believe strongly that the key to both is to manage knowledge systematically; their efforts to do so were started and nurtured to maturity by a strong leader; they give them adequate investment and priority; they bring to bear expertise and attention to detail; they ensure that knowledge activities are as rewarding for staff personally as they are for the organisation; they ensure that their wider culture and management practice support their management of knowledge; and they play a long game in knowledge initiatives, recognising that it can take a long time to get the practicalities right, for people to change their working habits, and for visible benefits to flow. We will consider most of these factors in more detail in the following two chapters. This chapter is concerned with just the first two: understanding how knowledge contributes to business strategy, and finding the conviction needed to drive initiatives through, over or round the obstacles that they will undoubtedly meet.

Frameworks for thinking

It has been a constant in business since the Industrial Revolution that firms need to improve their performance year by year simply to maintain their competitive position, unless they are lucky enough to enjoy a secure monopoly. Maintaining performance without improvement means slipping behind, and improving competitive position means improving faster than competitors. The fundamental case for knowledge management is that it is one of the most effective ways to accelerate performance improvement in current operations and the development of new products and services for tomorrow. But there are many aspects to improvement, and experience across many industries has shown that knowledge management programmes are apt to disappoint, and may fail entirely, if they are not driven by a clear view of what specific aspects of business performance they are intended to improve. Fuzzy objectives are confusing, and they encourage a scattergun approach that fires off ill-formed and under-resourced initiatives, overwhelms staff, and usually fails to make any real impact. Then disappointment turns to disillusionment, management enthusiasm and budgets wither, and knowledge management becomes just another discredited buzzword.

Reviewing the experience of over 70 companies, Booz Allen Hamilton found that disappointing results led to a 'disturbingly high' proportion of knowledge management programmes being cut back within two or three years. Two of the four principal problems they identified in less successful programmes were a lack of specific business objectives (there were only general aims, such as sharing best practices), and 'insufficient focus on one or two strategic priorities'. To bring programmes into focus it helps to have a framework for thinking about business strategy and relating it to knowledge objectives – a simple, structured way of differentiating between the aspects of performance in which the business needs to excel in order to achieve its aspirations, and those in which merely decent competence is enough.

Value disciplines

Business strategy is not an exact science, and there are many ways of thinking about it. One helpful framework was proposed by Michael Treacy and Fred Wiersema in their best-selling book *The Discipline of Market Leaders*. Drawing on an analysis of over 40 companies, they suggested that the most successful companies achieve their success by being leaders in one of three 'value disciplines': operational excellence, product leadership and customer intimacy.

- *Operational excellence* is leadership in price and customer convenience, achieved by minimising overhead, transaction and 'friction' costs, and optimising business processes; the business pay-off is market share. Firms like Dell, Amazon and Tesco work like this.
- *Product leadership* is based on the creation of a stream of state-of-the-art products and services by being creative,

commercialising ideas quickly, and relentlessly pursuing new solutions – if necessary, by making existing products obsolete. The pay-off is premium pricing. Intel, Sony, Canon, and Nike exemplify this approach.

- *Customer intimacy* involves tailoring products and services to meet customers' needs better and better, and personalising offerings to help customers achieve *their* ambitions. The pay-off is customer loyalty, which brings business stability, low sales costs, and opportunities to develop new products and markets. Management consultants and financial advisers often use this approach, and there are notable examples in other industries as diverse as logistics, telecoms and computing.

Treacy and Wiersema suggest that the best companies excel in one discipline, and are competitive – but not necessarily excellent – in the other two; only the most outstanding organisations are leaders in more than one. Although Treacy and Wiersema were concerned with commercial firms, the parallels in public service are obvious, and all three strategies are feasible options for professional services in both sectors.

Thinking deeply about a practice's value disciplines makes an excellent starting point for a knowledge strategy. Better knowledge management can raise performance in all three value disciplines, whether the aim is to build on an existing strength or to repair a weakness. But they call for different approaches to knowledge, both in the choice of tools and techniques and in the way they are implemented.

Operational excellence is based on doing routine things very efficiently, so the main objective of knowledge management in this case is to develop excellent, standardised processes that are as simple and foolproof as possible. In design practice, for example, this puts the emphasis on developing, recording, sharing and improving best practice in both design details and business processes to minimise costs *and* to make the client's experience happy and trouble-free, both pre- and post-construction.

Developing operational excellence requires rigorous, ongoing processes to discover what is working well and what less well – from the client's as well as the practice's point of view – and why; to spot mistakes, inefficiencies and successful innovations in individual projects; and to translate the lessons learned into process improvements. Techniques that can help do this include hindsight reviews, to analyse and understand project experience (ideally attended by all the organisations involved, including the client), in-depth client surveys carried out by independent interviewers, and benchmarking of performance against other practices and industries.

Processes and tools to share and record lessons learned and to support project delivery include bespoke, knowledge-rich IT systems that can provide timely prompts and checks throughout a project, standard documents and boilerplate text ready to adapt case by case, CAD libraries, efficient document storage and retrieval, and mentoring for new recruits.

Product leadership emphasises innovation, so there is less to gain from developing standard processes. The main aim of knowledge management in a product-leading company is to create conditions that encourage and support creativity, serendipity and lateral thinking. A rich and accessible resource of documented knowledge (from both internal and external sources) is important, but wide-ranging networks of personal contacts and – above all – talk and debate are even more so.

In professional practice, useful tools and techniques include software to facilitate networking, high-level mentoring to develop skills, foresight reviews to explore new ways to carry out projects, wikis to encourage the widest possible participation in developing the corporate knowledge base, and workspace design to encourage serendipitous overhearing, casual conversation and the development of trust.

Customer intimacy demands that management and staff have a deep understanding of customers, and earn their trust – in the case of design practices, not only of contractual clients but also of building occupants. The customer-intimate company looks far beyond the immediate objective of delivering a product that 'does what it says on the tin'; it seeks to bring its capabilities to bear in achieving the customer's wider ends, and invents ways to do this more effectively than the customer realised was possible.

Famously, IBM prospered for 30 years without either a price advantage or leading-edge products by analysing what its customers wanted to achieve and offering bespoke combinations of hardware and software that would deliver the business capabilities they needed, and more. Instead of expecting customers to have the IT expertise needed to specify a system and simply taking orders for boxes, IBM expected them only to know their own businesses, and sold them 'solutions' that wrapped the boxes in lucrative consultancy services. Equally famously, it failed to see that the standard desktop PC would offer enough capability for many purposes, and lost the new market to product-leaders such as HP and operationally excellent Dell. The customer-intimate company can no more afford to assume an unchanging world than any other. Today, IBM is successfully exploiting the complexities of IT–telecommunications convergence, and is prospering again.

Appropriate techniques for customer-intimate design practices include pre-project investigation of clients' business objectives, foresight and hindsight reviews involving both in-house staff and customers, post-project interviews with clients, and, in building design, post-occupancy evaluation including occupant surveys. Involving customers in reviews and giving them the opportunity to express their opinions and feel they are influencing the future not only builds their trust; it also makes it possible to learn important lessons about their real needs that would otherwise be missed.

Most knowledge activities necessarily have to be carried out by an organisation's own staff, but there are a few exceptions. Post-project interviews are one example. They are best carried out by independent consultants; feedback to staff is all too likely to be shaded by emotion (good or bad) and the unconscious wish to construct an acceptable memory and avoid compromising relationships. Misleading 'knowledge' can be worse than none at all.

Whoever collects it, context is often crucial in sharing customer knowledge effectively, so techniques that can preserve this, such as storytelling, learning histories and other forms of case study, can be particularly useful.

Making choices

The concept of value disciplines is useful because it helps to clarify objectives and focus attention on what matters most – in an entire business, as Treacy and Wiersema advocate, or simply in knowledge management. But focus means choice. How do you choose which value discipline to pursue? What are your strengths and weaknesses, and is it better to reinforce a strength or repair a weakness? Useful pointers can be found in:

- *Self-image and aspiration.* In owner-managed businesses, such as many professional practices, business strategy has to run with the grain of the owners' (and staff's) personal inclinations. If your real interest is in practising professional skills, product leadership is likely to be a more fruitful aspiration than operational excellence; if you really want to build a big business the opposite will be true. But strong interests tend to create weaknesses as well as strengths: the most innovative professionals are not necessarily the best at project delivery, for example. A realistic understanding of aspirations – and of any associated weaknesses – is one of the best starting points for deciding priorities in knowledge strategy.
- *Capabilities.* An effective business strategy is also shaped by capabilities. People – and teams of people – are simply better at some things than others, and it is good policy to play to your strengths. In an established business, years of recruiting to meet current needs (and which will inevitably

have favoured 'people like us') are likely to have developed a clear bias in one direction or another that can be difficult, expensive and risky to change. Knowledge initiatives can be used to reinforce these strengths, or – equally effectively – to repair the weaknesses. And when strategy does dictate a whole new direction for the business, knowledge initiatives can be powerful agents for change.

- *Public image.* This can be a surprisingly accurate reflection of capability, though it tends to lag changes, and has independent drivers of its own. If the realities of capability and public image conflict with aspiration, knowledge management can help convergence from both directions. In addition to their prime purpose of improving capability, demonstrably good processes for learning and sharing knowledge are already becoming a real marketing advantage with the more sophisticated clients.

- *Client base.* This can be an important consideration in strategy. Repeat business is often the lowest risk and most profitable, but a specialist niche may offer only limited opportunities for growth. People will pay a premium for product excellence only if they recognise it, so a professional practice that chooses this as its main value discipline needs sophisticated clients – and even then it may have to work hard to convince them that its innovations offer real benefits. Customer intimacy is possible only in a close and (at least potentially) long-term relationship, so it will not work with clients who habitually put every job out to tender. Inevitably, many of the clients for construction services simply want to buy a building in the same way that they buy a desk – with the minimum possible cost, aggravation and thought. The largest market will always be for operational excellence, and for a large practice this may inevitably be the dominant driver for strategy.

Of course, operational excellence, product leadership and customer intimacy are interlinked, and knowledge initiatives can often help more than one. Some of the techniques that help product innovation, for example, can equally well help business process innovation and operational excellence. Nevertheless, it is undoubtedly helpful to think about them – and often to take action on them – separately. If analysis of a practice's business suggests that more than one value discipline would benefit from better knowledge management, they can be tackled one at a time.

Other frameworks

Treacy and Wiersema's is not the only business strategy model that can be applied to knowledge management. Most organisations that are accustomed to using other models for their strategic planning will find that they can use them equally well for thinking about their organisational knowledge; they are all just aids to thinking, and they can be used almost interchangeably. Indeed, there are often clear similarities between them. To take just one example, consider the widely used balanced scorecard. In a KM context, its perspectives of *finance*, *customers*, *internal processes* and *learning and growth* translate into:

- *Financial perspective*: managing organisational knowledge and competence as resources, with initiatives to record knowledge of high business value, develop human knowledge resources (perhaps through mentoring, foresight and hindsight reviews and communities of practice), and develop tradable intellectual property such as patents and databases.
- *Customer perspective*: learning about customers' aspirations and needs through market research, post-occupancy surveys and client interviews.
- *Internal process perspective*: using knowledge to increase internal efficiency, with initiatives to make information quickly and reliably available (using state-of-the-art electronic tools for managing documents, information and communications), connect people (with people directories), capture best practice (perhaps with hindsight reviews), and support routine processes.
- *Learning and growth perspective*: initiatives to develop and encourage a 'learning culture', including such things as supportive staff appraisal metrics, workplace design that encourages informal interaction, hindsight reviews and links to research.

The similarity to Treacy and Wiersema's value disciplines is unmistakable.

Finding conviction

It is one thing to be persuaded intellectually that better knowledge management could pay dividends, quite another to *believe* that it is a strategic imperative. Initiatives driven by belief have a much higher probability of success; without it, they are apt to peter out, pushed aside by an endless succession of less important but more pressingly urgent demands on management time. The chief executive and other top managers need to become *convinced* that effective knowledge management is a top priority, and that the initiative they are about to launch is fully worth its cost.

It is impossible to do conventional ROI calculations for KM. Many of the costs are hidden, and the benefits are too intangible and uncertain. In practice, organisations today increasingly recognise knowledge management as a precondition for future success, and just do it. But investment decisions do not have to be based entirely on faith and hope. It is possible to put plausible bounds on the financial benefits of improved operational excellence, product leadership or customer intimacy.

All three translate into higher profits. If, for example, profits are 10% of turnover, then shaving 5% off costs by operational excellence – or achieving a 5% price premium through product leadership – will increase profitability by 50%. In other industries, a mere 5% increase in customer retention achieved by operational excellence has been found to increase profits by between 35% and 95% (it is not surprising that 'customer relationship management' has become such a hot topic!). Without a focus on customer intimacy IBM might well not have grown into the giant it is today, and Arup's reputation for innovation and technical excellence – in other words, product leadership – has undoubtedly been a major factor in its growth and success.

The profit potential of service improvements can be estimated by comparing, for example, the margins on prestige and routine projects or the cost of winning new projects from established and new clients, and then calculating the value of plausible changes in the mix. The results can be surprising. In a 50-person practice a benefit of just 2% of turnover would repay an investment of up to one person-year of effort or its financial equivalent – perhaps a mixture of consultancy, IT systems and staff time – within 12 months. When asked in knowledge management audits, staff in design practices typically estimate that they spend a third of their time reinventing wheels, doing rework, and looking for information. Saving 2% should not be hard. Good knowledge management can save far more, recouping the time spent in knowledge activities several times over, and effectively delivering the wider benefits for free.

Whatever business strategy is chosen, it is worth repeating that initiatives will succeed only if the underlying culture – in this context, people's innate sense of priorities and the forces that create it – works with rather than against knowledge activities. One of the commonest excuses that people offer for failure to engage in them is that they 'don't have the time', and that is simply a euphemism for 'I will be rewarded more for doing other things'. If that is what people feel, it is up to management at all levels to reorder staff priorities by changing both overt pressures such as personal appraisal criteria and time-booking procedures, and the implicit messages conveyed by their own behaviour; that is an essential part of the initiative. For learning and knowledge-sharing to become a habitual part of 'the way we do things round here' they need to be high priorities, both for the organisation and for its staff, and as valuable and rewarding at a personal level as they are strategically. Staff, as well as management, need to find conviction, and it is up to management to see that they do.

In the next chapter, we will consider the challenges of making major changes in the way an organisation manages knowledge – including introducing the concept of systematic knowledge management for the first time – and we will look more deeply into some of the other reasons why attempts have so often petered out at an early stage.

Chapter Four
The Challenges of Change

In the last chapter we looked at how business strategy can help to focus the objectives for knowledge management and get the best return from it. That is desirable, but not essential: even KM that is less than ideal can be worthwhile. However, as we saw in Chapter 1, many knowledge initiatives do not even achieve that. Booz Allen Hamilton's estimate that as few as one in six achieve 'very significant' business impact in their first two years, half achieve only 'small but important' benefits, and a third fail entirely is consistent with my own observations. Many start well, but peter out. It is a good discipline to aim for optimum alignment with business strategy, but the biggest challenge for firms setting out on knowledge management for the first time, or seeking to make major improvements in existing practices, is simply to avoid being among the third who fail, or the half who are probably disappointed.

In this chapter we consider why many organisations find it so much more difficult to establish effective knowledge management than they expect. The process only begins when senior management acknowledge that knowledge management is strategically important, so that hurdle has apparently already been overcome; there is rarely any overt opposition (there may even be widespread support); the ideas, procedures and IT tools involved are reasonably straightforward; and the cost is easily affordable. Why is it so difficult for managers to make knowledge part of their everyday thinking, and create truly learning organisations? What goes wrong, and how can the pitfalls be avoided?

Why initiatives fail

The causes of disappointment and failure are many, various, and impossible to disentangle fully.

Many of the early failures were the result of mistaking simple data and document management for knowledge management, and of buying complex, IT-based 'solutions' at high cost. Often, these were introduced much as an upgrade to a phone system would be, with no involvement from top management after the initial purchase decision, and with little consideration for the realities of how people work or what they might find helpful, let alone any engagement with knowledge in the true sense. The 'KM means IT' approach has now been largely rejected by knowledge management theorists and practitioners (if not by the software industry!), and the emphasis today is on people-centred techniques, understanding and meeting real business and knowledge needs, and fitting in with organisational culture. IT is an invaluable enabler, but it is a supporting actor, not the star.

Other causes identified by previous observers include failure to involve users in the design of tools and processes; the imposition of standard processes and tools on users with different and incompatible needs; knowledge bases full of information that is trivial or irrelevant; content that is too hard to interpret or use; inadequate quality control; inadequate technology; a belief that knowledge is power; and failure to demonstrate corporate benefits or maintain corporate support.

I have no doubt that these have all been significant contributors to the demise of knowledge initiatives in the past, and still are; several of them feature elsewhere in these pages. But most of them are technical pitfalls, in the sense that, once understood, they can be avoided (or at least minimised) without serious difficulty. All it requires is that the few people involved in developing an organisation's knowledge strategy, and in designing processes, tools and guidelines for their use, have enough expertise and put in enough effort. Much of this book is devoted to explaining pitfalls like these and how to avoid them. We have the benefit today of the pioneers' mistakes, and with good advice readily available from the literature and specialist consultants it seems to me that, though the dangers remain as real as ever, they need no longer be serious risks. It is just a matter of engineering around them.

There are, though, other difficulties that are more recalcitrant because they are rooted deeper in human psychology, and because it takes the combined efforts of many people to overcome them. I suspect they have always been among the major reasons for disappointing outcomes from knowledge initiatives (albeit often masked by more obvious, technical ones), and my observations suggest that they are the dominant ones today in professional services organisations. I have seen ten in particular recur again and again:

- Mismatch between expressed and felt beliefs
- Failure to form a powerful vision
- Failure to communicate the vision
- Failure to find early adopters

- Lack of urgency
- Failure to ring-fence necessary staff time
- Lack of stamina
- Lack of pervasiveness
- Poor tactical alignment
- Failure to understand the user's point of view and keep it in mind.

These can all be traced back largely to shortcomings in top-level leadership, and (as we shall see in the next chapter) that is where most of the solutions lie. Strong, inspirational, and sustained leadership is fundamental to the success of knowledge initiatives and the creation of a learning culture in professional services organisations; without it, the benefits are likely to be patchy and limited.

The failure of knowledge initiatives has often been attributed to lack of business alignment – to letting knowledge management develop as yet another bureaucratic function, without a clear relationship to business outcomes. As we saw in the last chapter, knowledge management's relationship to business strategy is certainly a major influence on the benefits that KM delivers, and thinking about it can help to bring the case for knowledge management into clearer focus and set appropriate objectives and priorities for it. Lack of alignment, though, seems to me to be a less significant reason for the outright *failure* of knowledge initiatives in professional services than it has been found to be in other industries, at least in the private sector. There are several obvious reasons why this might be so. All three of the fundamental processes that knowledge management supports are important to professional practices, albeit to degrees that vary from one to another. Professional practices are also unusually homogeneous organisations, both functionally and intellectually. They are typically managed by people who are also the owners of the business and remain involved in all its day-to-day activities, there is relatively little division of function, and staff at all levels tend to have similar educational backgrounds and share similar aspirations (so there is less need to articulate them). Finally, professional practices are rarely big enough to support even a specialist knowledge manager post, let alone a self-serving bureaucracy. However, business alignment occurs naturally when there is clear vision, success in winning hearts and minds, and good tactical alignment, so although it is not discussed explicitly in this chapter it is not ignored.

Mismatch between expressed and felt beliefs

Most top managers in industry and the public sector today express a belief in the value of managing knowledge and a commitment to making it happen. At a certain level this is genuine: it is, after all, self-evidently sensible to use knowledge better, and the value of knowledge and knowledge management is proclaimed by management gurus, business leaders and governments alike. The principle can be, and apparently often is, accepted more or less without thought – indeed, it would be difficult to dissent.

But there is a downside in unthinking acceptance of a principle: the lack of thought can mean that it has few connections with the deeply felt beliefs (conscious and unconscious) that actually drive behaviour. The elephant's rider knows what he wants to do, but the beast may have different ideas. Under the pressure of a typical manager's day, new ideas accepted unthinkingly are easily forgotten, and habitual priorities and ways of coping with problems, demands for attention, project delivery and budgets continue to rule.

We all know people who worry vocally about climate change, but still leave lights on when they walk out of a room, boil full kettles to make one cup of coffee, and drive to a newsagent ten minutes' walk away for the Sunday paper. The disconnection between high-level principle and everyday action in that case is just the same. It can be seen in so many aspects of behaviour that economists find it useful to distinguish between 'expressed' and 'revealed' preferences and beliefs, between what people's words say and what their actions say. Expressed beliefs are notoriously poor predictors of future behaviour, and for knowledge management to succeed, belief in it – especially leaders' and managers' belief – needs to go deeper than mere verbal expression. Learning and knowledge-sharing need to become part of their mindset.

Failure to form a powerful vision

Making a success of knowledge management – just like making a success of a merger or entry into a new market – requires a multitude of correct decisions to be made and correct actions taken, big and small, by many people in many different positions in an organisation. It would be impossible for anyone to remember a comprehensive set of decision rules, even if these could be formulated (and clearly they can't). In practice, most of our actions are guided by a relatively small number of deeply internalised ideas that we use to help us think about things – sometimes consciously and at length, more often unconsciously and almost instantaneously. The most influential of these – powerful principles, mental models of how the world works, and visions of the future – have widespread implications, most of which become evident to us only when we try to apply them. They underlie all religions, political movements, great scientific theories such as evolution, and on a more mundane level many aspects of our professional and personal behaviour. Their power to generate guidance for action whenever we need it eliminates the need for a library of case-by-case decision rules. Knowledge management needs to be guided by generative ideas such as these, encapsulated in a vision that shows how it can contribute to the organisation's future success.

A good vision inspires as well as guides. To succeed, it needs to present an attractive future briefly, vividly and memorably, in terms that people can relate to their own ambitions and everyday lives. It is not a statement of mission, philosophy, values, objectives or strategy (though it may refer to those in passing). Rather, it sets out what the organisation should be like in relevant respects several years in the future, why this would be good, and how the initiative in question will contribute to it. It is expressed in concrete, business-related and organisation-specific language to enable people to relate it to their own work and see personal as well as corporate benefits. It avoids bland universalities, but does not prescribe specific mechanisms. It carries an emotional charge, and evokes mental images. It enthuses people to think about it and start the process of internalising the ideas it embodies. And though it contains no details, it is pregnant with implications for future action, enabling people to work out the details for themselves.

There is a story that Steve Ballmer nearly left Microsoft shortly after Bill Gates recruited him as the company's first business manager. He decided he didn't want to be a bean counter. But Gates took him out to dinner and set out his vision for the company: he wanted to put a computer on every desk and in every home. That persuaded Ballmer, and he stayed on, eventually to become CEO. The vision for knowledge management in a professional practice needs to have something of that simplicity, concreteness and persuasive power. It needs to create an image in hearers' minds. It might seem difficult to achieve, but it must not seem impossible. It might talk, for example, about working together instead of like a set of separate, small practices; about releasing more time from reinventing wheels for more creative work; or about winning more interesting projects and becoming a more exciting and rewarding place to work.

For any complex management initiative to succeed, people – and particularly managers – throughout all relevant parts of the organisation need to be motivated to pull in the same direction, and be able to work out what that implies for their own work. A powerful vision is the basic tool for achieving both of those things, and without it initiatives tend to lose focus, lose coherence and peter out.

Failure to communicate the vision

In all fields, it is the job of the leader to communicate the vision. Not just to tell people about it, but to ensure that they understand it and commit to it; to win hearts and minds. Because effective knowledge management involves everyone in an organisation, the prime responsibility for this rests with the chief executive (in professional practices usually the managing director or managing partner), because he is the only person with a sufficiently wide span of authority. It means, too, that the chief executive needs to be actively involved in developing the vision, even if the underlying ideas come from elsewhere; if he has not become personally committed to it, internalised it and understood its implications himself he will not be able to communicate it with conviction, and it will not inform his behaviour and decisions as it needs to.

Vision is best communicated in person, face to face. That is the only way to ensure that it gets due attention, and to show the commitment behind it. Face-to-face communication takes people away from the immediate distractions of their work, allows them to concentrate, and gives them time to think about the message, if only briefly. And, crucially, it conveys the strength of commitment more vividly and convincingly than any other medium. However attractive change may seem to its instigators, some people always see threats in it. Apparently innocuous ideas such as learning lessons and sharing knowledge can seem threatening to those who see knowledge as a source of status and power, or fear erosion of their freedom to do things their way without critical scrutiny. Even the well-disposed need to be motivated: change requires positive effort, and it is always harder than no change. Any sign of lack of commitment will quickly be noticed, and people will sense that they can get away with foot-dragging and token compliance. That is fatal, and delegating communication of the vision to a memo (even over the chief executive's signature), a feature in the company newsletter or a colleague is likely to leave initiatives stillborn.

Chief executives cannot, of course, engage one to one with everyone in an organisation, and company-wide visions need to be interpreted in a variety of ways to transform them into concrete plans and practical action. It is too much to expect staff at lower levels and in specialist functions to make the imaginative leap from the big picture to detailed implications for their own work – besides, they might leap in different and incompatible directions. As well as the chief executive speaking and making his personal commitment visible to everyone, other directors and managers therefore need to repeat the process on a smaller scale, reinterpreting the vision in terms of their own areas of responsibility, progressively adding more detail to the big picture in a coherent way, and giving it equally visible support. In a very real sense, *they* make the practice's real strategy – the strategy that actually gets implemented – through their everyday decisions. They have to make the vision their own.

To do this effectively, they too need to engage creatively with it, think through its implications, and internalise it, and some at least need to develop their own emotional commitment to it. Enthusiasm will inevitably vary, and experience suggests that it is a good tactic to concentrate effort in parts of the organisation where the whole management chain buys into the vision. Outright opposition is rare, but passive resisters and foot-draggers at any level can block progress. They need to be sidelined, and management champions who can overcome local obstacles need to be given unstinting support. Laggards will follow when they realise what they are missing.

Momentum needs to be maintained, too. Backsliding is all too easy, and ideas that seem perfectly clear in early presentations and conversations can quickly lose focus as time passes and they are overlaid by the urgencies of day-to-day practice. There is no shortage of excuses. 'Yes, learning more and using knowledge better are vital to the future of the business, but I'm too busy today, tomorrow will do; yes, we want everyone to contribute to the knowledge base, but I'd really feel more comfortable if everything was checked by an expert; I know it will help buy-in if we let the staff decide what to include on their personal pages, but *I* want it this way; *I* know what went wrong with this project, so it isn't worth spending time on a review.' Without continuing reinforcement from the chief executive and his allies the results is likely to be a disappointing knowledge system that delivers much less than it could have done.

Failure to find early adopters

Momentum also requires a steady accumulation of solid achievement, and that needs early adopters prepared to wrestle with unfamiliar tools and techniques and make them work. Producers and consumers are equally necessary in a knowledge economy, but production – organising project reviews and community of practice events, writing knowledge base articles and so on – has to come first. Fortunately, only a few tools, such as social networking tools, need large numbers of contributors in order to become useful. Mentoring, foresight and hindsight reviews, project directories and even communities of practice need only a few activists to start generating worthwhile benefits. Indeed, experience shows that even well-established knowledge systems are usually maintained by an active minority while the majority remain minimally contributing consumers. Activists are invaluable, and early-adopter activists particularly so.

Some of these are simply enthused by new technology and new possibilities, and curious to find out what it can do. They are the classic early adopters who were happy to pay high prices for the first iPods. Others are people with a strongly felt need, to which the new knowledge systems offer a solution. Often, this stems from frustration: seeing mistakes repeated or having to spend hours collating information for bids, for example. Both types can help, but needs-driven activists are more valuable than technology enthusiasts. They are motivated by business benefits, and their successes are more likely to be persuasive exemplars.

In organisations where managers are also working professionals the management champions may also be the first activist adopters. They are ideally placed to take the lead in activities closely linked to business operations, such as hindsight reviews and project directories. It can be counterproductive, though, for those that depend on enthusiasm for a technical topic to be management-led. Successful communities of practice, for example, derive much of their vibrancy from their independence of management and their detachment from immediate job pressures. Similarly, it is important to demonstrate as quickly as possible that it is knowledge rather than formal status that qualifies people to contribute to a knowledge base. Coincidences of enthusiasm and opportunity, wherever they occur, need to be sought actively and exploited to the full in order to get new knowledge systems off to a good start.

Lack of urgency

It is a truism that the urgent displaces the important, and there are few corporate activities both more important and less urgent than knowledge management. Knowledge really is the key to future business success, but there is rarely any obvious penalty for putting off a knowledge management initiative – or, indeed, any individual knowledge activity – until tomorrow. And, of course, though we are too busy today we all expect to have more time tomorrow, or next week; we will be able to do it better then. The trouble is that, as we all know, one delay has a habit of leading to another.

The belief that there will be fewer distractions and more time in the future is just a comforting illusion. Recent psychological research has shown that humans are systematically biased towards overconfidence, particularly about the future and about our ability to control events. In fact, tomorrow is likely to be filled with just as many urgent tasks as today. And although routine knowledge activities usually have only modest individual value, the cumulative loss from delaying them all, for any significant time, is potentially very large. When initiatives such as the introduction of a social networking tool, project hindsight reviews or a whole knowledge management programme are allowed to drift, everything that they would enable gets delayed, and some opportunities are lost for ever. Worse, long postponement inevitably means that the vision dims, enthusiasm leaks away, and belief in the importance of knowledge management is progressively eroded. We've survived without it, so how can it matter?

Psychologist Piers Steel of the University of Calgary has analysed data from over 500 studies of procrastination to look for its root causes. He concluded that there are four factors that consistently matter more than any others: how pleasant a task is; how confident people are of success; how easily distracted they are; and how long they expect to have to wait for the pay-off. This goes a long way towards explaining why knowledge management initiatives are so prone to delay in professional services: procrastination is almost inevitable when busy practitioners lack real belief in their importance, they are aware that benefits will take time to become visible, and they would much rather be doing other things. Without a strong sense of urgency to counter factors like these, transient distractions destroy momentum, and everyday pressures can become an excuse for recurrent delay, lowering of ambitions, and even complete abandonment.

Failure to ring-fence necessary staff time

When knowledge thinking and knowledge activities are part of the culture they make no net demand on staff time. They certainly take time, but they save even more by enabling people to work more efficiently and creatively. While knowledge systems are under development, though, they *do* make net demands on time, and the people whose contribution is most vital – managers and experts, at all levels – are often those under the most time pressure from their routine work. Without ring-fenced time for knowledge work the temptation to avoid it can be great, even in the absence of competition from self-evidently top priority tasks. For knowledge initiatives to succeed, the pressures that prevent people from playing their part need to be identified and reduced.

Time pressure is often a result of financial targets that focus on short-term performance at the expense of long-term business success. Time spent developing knowledge management appears as an immediate cost, whereas efficiency savings and business gains may not accrue until a later accounting period. They may even accrue in a different part of the company, leaving some business units with cost burdens even though the company is benefiting overall. Imbalances can persist even when knowledge management has become routine. Without an appropriate budget allowance, a manager with local profit responsibility needs to be unusually saintly to continue supporting knowledge activities in a situation like this.

Lack of stamina

The weakness of knowledge management in its competition with more urgent activities for attention and time leaves it continuously vulnerable. It is at constant risk of being sidelined until habit embeds knowledge thinking into everybody's mental fabric and daily work, and it becomes evident to all that the benefits outweigh the costs at both personal and corporate levels. Successful knowledge management depends on pervasive changes in the way people think and act, and these need first to be developed and then reinforced by practice, practice and more practice until they cease to need conscious effort and become habitual – simply 'how we do things round here'. Achieving this demands continuing encouragement, support and sometimes overt pressure throughout the management chain over a surprisingly long time after the vision is first developed – often several years.

Many boards have seen knowledge management as something that can simply be delegated to middle-ranking staff and left to its own devices. That is an unrealistic expectation, and a recipe for modest success at best. The chief executive and top management need to provide continuing reinforcement for the vision and sense of urgency well beyond the first announcement, give knowledge activities continuing practical support, and avoid creating new obstacles through other management decisions. If the vision is allowed to blur and become distorted, and urgency is allowed to fade, old priorities quickly reassert themselves and displace the new ones as the pressures of everyday work squeeze out the thought and attention needed to establish new ideas. They need to be refreshed time and again.

But even energetic advocacy loses its power in time if there is nothing else to help maintain commitment. Knowledge activities need to produce visible benefits as soon as possible, both for the company and for its staff. This can be achieved by introducing the most immediately rewarding first and taking advantage of favourable situations. Foresight or hindsight reviews of well-chosen projects can be a good choice, because participants enjoy them, and it is obvious when useful lessons have been learned; the harder task of sharing the lessons across the practice can be left until later. One or two fruitful reviews can be enough to convince business unit managers that reviews justify the time they take, and give them the motivation they need to do more. However, it is unlikely to persuade them to divert enough time from fee earning to develop systems that need much more up-front investment to become productive, such as knowledge bases. Financial support in forms such as reduced profit or utilisation targets may need to continue for some time to ensure that managers give staff time for knowledge activities and prevent short-term, local pressures from sabotaging long-term, practice-wide ambitions.

Finally, management at all levels needs to make sure that staff engage in knowledge activities often enough for them to become habits. They need to ensure that staff cannot forget or ignore them. This requires sustained publicity and encouragement, and it may mean making core knowledge activities such as completing personal and project records mandatory, and insisting that they are done.

Lack of pervasiveness

Size matters in knowledge management. Some techniques, such as the systematic review of experience to learn more from it, can be helpful even in the smallest practice, but many need a critical mass of users to become worthwhile. The potential benefits from all of them grow with the size of their active user base. It is not only headcount that matters. A knowledge base needs a critical mass of content to make it worth having at all, and as it grows further it becomes able to answer an increasing number of questions, worth consulting more often, and more likely to become a routine working tool (rather than a last resort). Further, the scrutiny of more users helps raise the quality of the content. Together, factors such as these make a knowledge base with twice as much well-targeted content more than twice as valuable. The same is true for social networking tools. The number of possible connections between people increases much faster than the number of people,[1] and, as Mark Granovetter pointed out in 'The strength of weak ties', information from outside people's immediate circle can be particularly valuable because it is more likely to be novel.

There can also be beneficial synergies between different knowledge management techniques, and between knowledge management and other corporate activities. Project hindsight reviews and communities of practice are both prime sources of content for a knowledge base, for example, and social networking tools can provide valuable source material for making up project teams and for HR activities such as staff appraisals.

To get the best from knowledge management, it therefore needs to be a pervasive activity, encompassing all the staff (or at least all the professional staff) and all the activities in a practice. Half-hearted, piecemeal and isolated initiatives are apt to disappoint.

Poor tactical alignment

There is recurrent reference to 'business alignment' and 'strategic alignment' (which are essentially the same thing) in the management literature. Tactical alignment is mentioned less, but it is equally important. Strategic alignment demands that the direction and priorities of major initiatives and investments be driven by high-level business needs and priorities – by corporate vision, aspirations, goals and objectives. In the knowledge management context this can mean, for example, giving priority to activities that can support a particular value discipline, such as foresight reviews to foster creative design and support product leadership, or hindsight reviews to increase quality and operational efficiency. Tactical alignment, on the other hand, aims to match the emphasis and design of knowledge systems to the organisation's specific, detailed and immediate operational needs, and to the practicalities of the way the organisation, and human

---

[1] In fact as $n(n - 1)/2$, so that there are three possible connections between three people ($3 \times 2/2$), six between four ($4 \times 3/2$), fifteen between six and so on – an increase almost as fast as a square law.

psychology, work – if necessary, adjusting existing procedures, rewards, pressures and IT systems to mesh better with knowledge activities. This can have just as profound an effect as strategic alignment on the business value that flows from knowledge initiatives.

Tactical alignment can take many forms. If an unusually large number of key staff are due to retire over the next few years, for example, it might mean setting up special mentoring arrangements to transfer their expertise to the next generation, or delaying the introduction of new systems until new staff are in post. Careful design of new systems to ensure that they support particularly time-consuming and important tasks such as preparing bids will help sell them to an important layer of management. Changing annual assessment criteria to include participation in knowledge activities can help show that it is valued, and make it more difficult for sceptical managers to discourage it. Ensuring that social networking tools can draw basic data from the personnel database gives the system immediate value, even before staff add any other information about themselves. On the other side of the coin, there is little point in the chief executive exhorting staff to put more time into voluntary knowledge activities if managers are judged solely on billable time; priorities reflect pressures, and when necessary the pressures need to be changed. It takes imagination and a wide knowledge of the organisation to recognise opportunities and conflicts like these, but it is usually clear what needs to be done to achieve alignment once they have been recognised. Resistance to changing established systems and practices needs to be overcome by strong leadership and negotiation.

It can be more difficult – but it is equally important – to make knowledge activities fit harmoniously with organisational culture and take adequate account of psychological factors. Culture is often said to be a key factor in learning and knowledge-sharing, and so it is. It is much easier to introduce new knowledge systems successfully in an organisation where personal initiative, spending time looking things up and chatting to colleagues are the norm, than in one with a headsdown, do what you're told culture. Unfortunately, changing the culture directly can be dauntingly hard. Unlike formal procedures and database software, culture is not designed. It simply emerges from practice, often reflecting beliefs deeply held by senior management, and it has a pervasive (if often unrecognised) influence on day-to-day management practice and staff behaviour. This can make it remarkably resistant to change. One way to make progress is to concentrate on changing habits; eventually culture will follow. The principal purpose of a social networking tool, for example, is to connect people with questions to people with answers, but it is worth making it an excellent internal phonebook as well in order to bring it into daily use, and then people will gradually start to use its more sophisticated facilities.

When the culture is particularly unfavourable it may be worth developing knowledge management in stages over several years, starting with low-conflict steps and accepting that the benefits will take longer to emerge. Whatever the context, understanding and accommodating the realities of organisational culture are prerequisites for success.

Human psychology is even harder to change than culture, and even small details can influence behaviour. Architects, for example, work with drawings and pictures all the time, and that makes them unusually sensitive to visual details. They typically have a strong preference for design-rich, Apple Mac-like interfaces that make extensive use of graphic icons, and even dislike of a colour can be a barrier to acceptance. Research has shown that apparently superficial factors like this can make a dramatic difference to success. In the 1960s Yale psychologist Howard Leventhal conducted an experiment with a booklet designed to persuade students to get a tetanus shot. To measure the effectiveness of fear as a motivator, he gave some students a version that emphasised the risks and included alarming photographs of tetanus victims, and others one that was couched in less alarming terms. Both versions offered free shots at the health centre. A follow-up survey showed that the group that had been given the more frightening version were indeed more convinced of the danger of tetanus and more likely to say that they intended to get inoculated. In the event, though, hardly anyone from either group actually visited the health centre in the following month. But then Leventhal repeated the experiment with one small change – the addition of a campus map with the health centre highlighted, and its opening times – and the inoculation rate leapt from 3% to 28%. Again, the fear factor made negligible difference to actual behaviour: the two versions of the revised booklet provoked almost identical responses. Leventhal hypothesised that, since it was unlikely that any of the students actually needed the map, the critical difference was that it connected the medical theory to their personal lives, and showed them what to do next. There are no simple rules for getting details like this right in knowledge systems, but experience in the field, sensitivity, care and good use of user feedback can all help avoid pitfalls and make initiatives more successful.

Poor tactical alignment of any kind reduces the returns from investment in knowledge management. It can lead to opportunities being missed, to conflict over priorities and resources, and even to complete failure. This is a further reason why knowledge management cannot be an independent, stand-alone activity, and it cannot be delegated to middle-rank staff who have only local knowledge and lack the authority to override sectional interests and judiciously coerce the unwilling. Business unit managers and other functions need to collaborate constructively in developing the strategy, systems, procedures and pressures. Without continuing, active involvement from top management, lack of engagement and turf wars can make effective tactical alignment impossible, and can compromise the whole knowledge management effort.

Failure to understand the user's point of view and keep it in mind

In knowledge management, as in business, the customer is king. The majority of learning and knowledge-sharing occurs in unmanaged, voluntary activities that happen only if people want them to. Knowledge activities have to be *sold* to their users, and the usual economic framework for discretionary spending (in this case spending of time and effort rather than money) applies: people will buy only if the benefits as *they* see them outweigh the costs. The business case may rest on knowledge's importance in securing the firm's future, but that does not interest grass-roots staff; they just want the job on their desk to be a little easier or more rewarding. Management can certainly provide enablers such as review workshops and knowledge bases, but its power to make people use them is strictly limited. Crude coercion, without also making the activity a reasonably attractive buy, is likely to lead to token compliance – making the unavoidable minimum contribution to a review workshop, or even visiting a knowledge base simply to register an access in the server log.

It is vital to take the user's point of view into account in the detailed design of knowledge tools and processes. Poor user interfaces, for example, have been the downfall of many IT-based systems. One university research group of my acquaintance decided to use a wiki as the framework for a knowledge base, but rather than invest effort in installing a proper wiki package chose to use existing software that superficially appeared to provide comparable facilities. Unfortunately it lacked the user-friendliness of a true wiki, and when the first would-be users (including me) – full of enthusiasm – tried to upload documents they were unable to discover how. User complaints elicited an explanatory email a few days later, but by then it was too late: the enthusiasm had evaporated, we were all busy with other things, and the initiative failed. Details can be disproportionately important.

There are particular dangers when existing systems and resources are simply relabelled without considering whether they need to be adapted to suit a new role or user base. Long-established staff databases designed for managers and HR personnel, for example, are not infrequently relabelled 'yellow pages' with only a cursory attempt to fit them for a wider role as social networking tools. Similarly, superficial project reviews involving only in-house project leaders may be relabelled 'hindsight reviews' without bringing in other parties or making them systematic examinations of what went right, what went wrong, and why. Existing systems and procedures designed with other purposes in mind are rarely fit for purpose as they stand, in a knowledge management context.

Pitfalls like these can be avoided by good initial decisions about knowledge strategy, and by good design of processes, systems and guidelines for their use. However, their subsequent success often turns on factors that are outside the original designers' control, and may be peripheral from a strictly knowledge perspective. It is just as important for the managers and other people who have ongoing responsibilities for knowledge activities to have the user's point of view in mind: it needs, in fact, to become a habitual way of thinking. A senior manager may want to hold hindsight reviews before or after normal office hours in order to avoid encroaching on his packed working day, but staff are likely to see this as an imposition – especially if it conflicts with their commuting arrangements or regular social commitments. The business value of social networking tools may lie in details of knowledge and experience, but staff in general are more likely to use them if they also help to find partners for a tour of real ale pubs. Requiring contributions to a knowledge base to be submitted through an expert moderator may seem a necessity to ensure quality, but experience shows that it is highly inhibiting, and a mass of somewhat variable content is far preferable to a quality-controlled but almost empty knowledge base. (It is unnecessary, too: in a well-used knowledge base users quickly correct errors, and in any event most knowledge is shared in quality-uncontrolled conversations between colleagues.)

It is tempting to use whatever material happens to be readily available to start the ball rolling in a new knowledge base, but if that is not what users want, failure is an almost inevitable consequence. As we saw in the last chapter, information becomes useful only when it completes a knowledge jigsaw and enables someone to do something they could not do before, or to do it better. Information resources such as knowledge bases stand or fall by their ability to complete users' jigsaws. Content that is irrelevant or inappropriately configured – which is suitable for just-in-case but not for just-in-time use, for example – simply clutters up the system and puts users off.

Managers' failure to anticipate and accommodate the user's point of view in respects like these has often caused knowledge initiatives to atrophy. This blind-spot is not specific to knowledge activities. As Harvard professor Teresa Amabile has put it, 'most managers are not in tune with the inner work lives of their employees.' This needs to change in the would-be learning organisation. The user's view is crucial in every aspect of knowledge management, and it is not enough to take it into account in the early stages of developing a vision or a strategy, or designing tools and processes; it needs to be a permanent influence.

Difficulty is normal

With so many pitfalls it is tempting to conclude that introducing knowledge management is an especially difficult process, and professional practices an especially unfavourable business context for it. Not so. The analysis from which these ten pitfalls emerged was based on evidence from a relatively small number of organisations, mostly in professional services. But corporate change in general – of which the development of effective knowledge management is an example – has been studied extensively, and the results are surprisingly similar across a range of industries. Research on over 100 companies led Harvard Business School Professor of Leadership John Kotter to identify eight chief causes of failure in change efforts ranging from the introduction of total quality management to mergers and acquisitions:

- Allowing too much complacency (failing to establish a sense of urgency)
- Failing to create a sufficiently powerful guiding coalition
- Underestimating the power of vision
- Undercommunicating the vision
- Permitting obstacles to block the vision
- Failing to create short-term wins
- Declaring victory too soon
- Neglecting to anchor changes firmly in the corporate culture.

Of these, Kotter considers four to be the source of most failures:

- *Failing to establish a sense of urgency*, which he says is at the root of over half of all failed change efforts.
- *Undercommunicating the change vision.* Most leaders, he says, under-communicate 'by a factor of 10'.
- *Declaring victory before the war is over.* Kidding yourself about the difficulty or duration of organisational transformation, he says, can be catastrophic: one third of the way into the process you may see only a tenth of the possible results, and if you settle for too little too soon you will probably lose it all.
- *Failing to build a guiding coalition.* The people to watch are those with most to lose from change, often just a step or two below the chief executive. And though many people want to believe in the chief executive's vision, their managers often give them reasons not to. People with the right commitment and capabilities from all levels in the organisation need to be brought together in a group empowered to overcome any opposition.

The correspondence between Kotter's eight chief causes of failure in change programmes in general and the ten factors that have compromised the success of knowledge management initiatives in professional practices is evident. 'Allowing too much complacency' is essentially the same sentiment as 'lack of urgency', 'underestimating the power of vision' the same as 'failure to form a powerful vision', and 'undercommunicating the vision' the same as 'failure to communicate the vision'. 'Permitting obstacles to block the vision' is an aspect of 'poor tactical alignment', and 'failure to find early adopters' an aspect of 'failing to create a sufficiently powerful guiding coalition'. 'Failing to create short-term wins', 'neglecting to anchor changes firmly in the corporate culture' and 'declaring victory too soon' are all aspects of 'lack of stamina'.

In this respect, at least, professional practices seem to be indistinguishable from other kinds of business, and the introduction of knowledge management indistinguishable from other change processes. If this is so, another of Kotter's conclusions is salutary: that fewer than 15% of the companies he studied transformed themselves successfully – a figure remarkably close to Booz Allen Hamilton's estimate that only one in six knowledge management initiatives are fully successful. The pitfalls are very real – but provided they are recognised and treated with due seriousness, much can be done to avoid them.

Booz Allen Hamilton's perspective on causes of failure is again broadly similar to those we have considered in this chapter. They identified four principal problems:

- *Lack of specific business objectives*, with only general aspirations such as sharing best practices or stimulating collaboration – a problem that, as we have noted, seems to be less common in professional practices than elsewhere
- *'Incomplete program architecture'* – broadly similar to lack of pervasiveness
- *Insufficient focus on 'one or two strategic priorities'* – which leads to failure to create quick and visible wins
- *Lack of 'active, ongoing involvement' from top management* – a common factor in failure to communicate the vision and ring-fence necessary staff time, and in lack of stamina.

Booz Allen Hamilton hypothesised that these all stemmed from 'top management's failure to play its accustomed roles of leadership and management'. Exactly so. In the next chapter we will consider what that role entails in a knowledge management context.

Chapter Five
Leadership and Other Roles

The common thread that runs through most analyses of corporate change is the vital importance of leadership, first by the chief executive and then by other directors and managers as they take up the baton and carry it through the organisation. Mismatch between expressed and felt beliefs, failure to form a clear vision, failure to communicate the vision, lack of urgency, lack of stamina and many failures of tactical alignment – and the pitfalls on Kotter's list – all have a large 'hearts and minds' component, the province of leadership.

Action starts where the buck stops

Kotter argues that most change programmes put too much stress on 'data gathering, analysis, report writing and presentations' and too little on 'the feelings that motivate action' – in Jonathan Haidt's terms, too much stress on the rational rider and too little on the wilful elephant. This chimes with conclusions reached by Stanford professors Jeffrey Pfeffer and Robert Sutton. They were struck by the apparently small impact that the mountain of books, articles, training programmes and conferences on management have on what managers actually do. People, they observed, 'obviously knew what to do, but didn't do it', and surveys showed that many managers were themselves aware that things they knew their company should be doing were not in fact being done. In a vivid phrase, Pfeffer and Sutton called this discrepancy the 'knowing–doing gap'. Surveys and study of real-life cases led them to identify five barriers to turning knowledge into action:

- The tendency to treat talking about something as equivalent to actually doing something about it: spending time taking carefully considered decisions, making presentations, writing

and discussing reports, planning, and issuing mission statements instead of getting on with simple actions

- Relying on memory, precedent and conventional wisdom
  even when it was evident that changed circumstances made
  them invalid
- Fear and distrust inhibiting innovation, risk-taking, honesty,
  and questioning of authority
- A focus on the measurable to the exclusion of judgement,
  leading to the neglect of crucial non-measurable factors and
  distortion of strategy and priorities
- Internal competition turning friends into enemies, and
  undermining teamwork and knowledge-sharing.

Again, these are essentially emotional barriers – further examples of
mismatches between observations and expressed beliefs (conscious
knowledge) and felt beliefs (the unconscious beliefs that actually
determine behaviour), between the intentions of the rider and the
inclinations of the elephant.

All this research – Kotter's, and Pfeffer and Sutton's, on industry in
general, and the work with professional practices that gave rise to this
book – puts the prime responsibility for success in developing knowledge management and creating a learning organisation squarely on
the shoulders of the chief executive. The buck stops on his desk. If he
gives out the wrong emotional messages, no words or plans can repair
the damage. The better he plays his part the better the end result is
likely to be. And his first task is to examine his own past actions, discover his own revealed beliefs, and if necessary convince himself to
change them. This calls for courage and honesty.

It is unlikely to be coincidence that BP, the company that has been
most visibly successful in exploiting knowledge management in the
UK, was led for over a decade by one of its highest-profile and most
passionate and articulate advocates, Lord Browne. In the 1980s BP
was a second-tier player in the oil business, with declining reserves,
among the highest development costs in the industry, and an uncertain future. Its renaissance as an admired and successful first-tier
company is often credited to Browne's leadership, and to his visionary
insistence on the central importance of knowledge and learning in
modern business. He not only invested financially in knowledge management but also backed it with personal example and sustained
advocacy and support, and the results speak for themselves. Browne
makes a stark contrast to his predecessor but one, Robert Horton.
Horton said the right things, proclaiming that the company's values
must include openness, care, teamwork, empowerment and trust. But
his words were fatally contradicted by a policy of downsizing and
cutting capital spending, and a management style widely regarded as

abrasive. The result was fear, cynicism, defensiveness and distrust, and he was replaced after two years despite the undoubted value of his organisational reforms.

In a public company such as BP it is not hard for a committed and forceful chief executive to drive through an initiative that is low cost (in relation to profits) and risk free. The leadership task can be more challenging in a business privately owned by working directors or partners, as many professional practices are. The chief executive's positional authority is less, and the fragmentation of power places a premium on his ability to communicate his vision persuasively to his co-owners and build an alliance strong enough to make it a reality. It adds to the challenge that an owner-managed practice often operates less like a unified company and more like a loose federation of separate businesses, each with its own different culture, ambitions and priorities. A further complication in many practices is that a high proportion of staff (even relatively junior ones) have a considerable degree of personal autonomy, and the nature of their work makes it less susceptible to detailed management than in most other organisations. Even at the grass roots, the success of change programmes can depend more on people buying into the vision than on managerial fiat.

Engineering consultancy Arup is an organisation in which people have done just that. Founder Over Arup embedded a respect for learning and knowledge in the consultancy's DNA from the start, speaking over 60 years ago about how difficult it was even then for engineers to 'become familiar with the complete range of modern technical possibilities', and of the need for design practices to develop a 'composite mind', sharing knowledge across the organisation. Knowledge has been part of the Arup mindset ever since, supported by an evolving range of activities, processes and software tools, and the practice has reaped the benefit of its founder's vision by maintaining a premium position in its market.

However, visionary leadership alone is not enough to ensure the success of a knowledge management programme. Practical action, tactical alignment, accommodating the staff's point of view, and getting the detailed design of procedures and IT tools right require expert knowledge and understanding of knowledge management itself, the cultural context, working practices, and the IT and other systems that need to work together.

Practical leadership

There are several practical steps that company leaders can take to give initiatives the best possible chance of success.

Use debate to sort out your beliefs and develop your understanding

It is impossible to change one's own deeply felt beliefs by effort of will, or change other people's by managerial instruction. The brain can only be rewired by spending time in serious thought, testing new ideas – and especially, in this context, the idea that knowledge management really *is* vital to continuing corporate success – against established beliefs, personal experience and external evidence, and thinking through some of the personal implications of the new ideas.

One good approach is to get together with a small group of like-minded colleagues for a series of informal debates, ideally also including someone (from outside, if necessary) who is familiar with knowledge management concepts and practice. Simply having to articulate thoughts helps greatly to clarify and develop them: there is more than a grain of truth in the saying 'How can I know what I think until I hear what I say?'. Exposure to other people's thinking is equally valuable: their different perspectives can contribute valuable ideas, point out additional pitfalls, and help to develop arguments strong enough to persuade the unenthusiastic and to neutralise active opposition.

Debates like this help to clarify new understanding and connect it to – and if necessary change – pre-existing assumptions and beliefs to a degree that is difficult to achieve by solitary thinking. Actively using knowledge is a uniquely powerful way to develop it further and anchor it in the mind. It would be hard, for example, to develop a worthwhile knowledge of algebra without solving equations, or of history without writing essays. Debates serve other purposes as well. They provide the foundations for a clear vision, prepare for communicating it more widely, and help to build the guiding coalition that Kotter advocates.

Ring-fence time

A final advantage of debates is that they ring-fence time for thinking: timetabling a meeting is one of the best ways to avoid interruptions. Chief executives and managers need to continue to ring-fence time for knowledge management activities until they become habitual for everyone. As we have seen, commitment to the principles is only the first step on a long road. Senior staff need to create time and the conditions for mental engagement and leadership, even if they delegate the design and implementation of knowledge tools and techniques to others. Confronting personal beliefs, developing a vision, communicating it, building alliances, setting an example, keeping an eye on progress, and occasional intervention all need time, and they are all essential.

Less important work must be delegated, if necessary, to free up sufficient time. Any difficulty in identifying activities that are less important than creating a learning organisation is a hint that expressed and felt beliefs are still mismatched: back to square one.

Get help

Equity partners and shareholding directors in professional practices are characteristically hands-on, often combining directorial responsibilities with personal involvement in day-to-day management, marketing and continuing fee-earning. This can make it particularly difficult for them to find adequate time to drive knowledge management initiatives, and the result is often delay after delay. It is always tempting to believe that when the current crisis has passed there will be time available, but this rarely happens – another crisis just appears to replace it. The solution is to identify the core leadership responsibilities that *must* be discharged personally, and get help with everything else.

Knowledge management poses unusual staffing problems, which we consider in detail later in this chapter. In a true learning organisation responsibility for managing knowledge activities is widely diffused. It is a key part of *every* manager's role, and although certain roles such as leading communities of practice fall naturally on domain experts, these are usually best rotated between several people to spread the load. In a large organisation it may be appropriate to create a full-time 'chief knowledge officer' post to support managers, CoP leaders and others by providing specialist expertise, practical help, encouragement, coordination and oversight for knowledge management activities. However, only the very largest organisations are likely to be able to recruit a sufficiently high-calibre and experienced specialist to guide the initial creation of the system; the subsequent role has too little to offer in smaller organisations. By the same token it is rarely satisfactory to give an existing member of staff responsibility for creating a knowledge system, whether on a full or (even worse) a part-time basis.

The best source of help is often an outside consultant. Consultants can contribute expertise, insights, detachment, focused attention and time unlikely to be available in any other way. They can also relate to the board and to staff at all levels without the complicating baggage of an insider or the inhibitions of a formal status in the company. They can reduce misunderstanding, help clear thinking, and ensure that key issues are not overlooked, and they are much more likely than colleagues to ask the challenging questions that always need to be asked. Consultants' independence and detachment make them better equipped than in-house staff to carry out knowledge audits and discover how staff really perceive the company's culture and management style, how well knowledge systems are really working, and what clients really think about the practice's work. In the early stages of a knowledge initiative they can give invaluable help in developing and communicating a vision, developing overall strategy, designing tools and processes, guiding practical implementation, monitoring progress, and coaching in-house staff in knowledge roles. Later – as in other areas where high-level specialist expertise is desirable, but where it would be inappropriate to create a full-time post – it can be invaluable to have continuing access to outside expertise through a non-executive directorship or a similar arrangement.

Lead by example

The most persuasive evidence chief executives and other managers can give that they believe in the importance of knowledge is by visibly practising what they preach. It is worth the effort for that reason alone for them to be among the first to complete personal pages, to contribute to the knowledge base (after all, in a professional practice the chief executive is usually one of the top experts), to participate in foresight and hindsight reviews, and perhaps even to take a short turn as a knowledge base moderator. Practical engagement as a contributor and user has another benefit, too: without personal experience of knowledge systems and activities, managers are ill equipped to help guide their future evolution and ensure that they are used to best advantage.

Senior management can also give a lead by showing that they use knowledge systems routinely to support their own work. Foresight, hindsight, knowledge bases and mentoring in particular can serve business management just as well as professional ends. Hindsight reviews, for example, can help in learning lessons from any discrete activity such as a major bid or opening a new office. Mentoring is an ideal way for managers who are changing role or retiring to pass on their knowledge to their successors. And a knowledge base is just as good a repository for important marketing and management knowledge as it is for more technical knowledge.

Create motivations and remove demotivations and obstacles

The success of knowledge management depends largely on people at the grass roots engaging in discretionary knowledge activities – doing things that they do not *have* to do. It follows that management can *make* learning and knowledge-sharing happen only to a limited degree: beyond that, managers can only create a conducive environment by providing enabling systems and motivating staff to use them. There used to be a widespread expectation that simply providing the systems would be enough; they would be so self-evidently valuable that users would flock to them. However, faith in the 'build it and they will come' approach crumbled as evidence accumulated that they don't, unless the benefits are self-evidently overwhelming – as, for example, they are with tools such as Google. Motivation is crucial, because people will not act without it. And belief in the strategic value of knowledge as such is enough to motivate only a visionary few; most people have more pressing concerns and need something more tangible.

It is one of the key challenges in creating a learning organisation that few knowledge activities are immediately rewarding in themselves. Completing personal pages, for example, is simply a chore for early contributors because there are too few other entries for the system to be useful as a source of contacts; it costs time, and the system offers little beyond a phone book in return. As personal pages continue to accumulate, however, social networking tools gradually become more useful, and eventually a tipping point is reached where they can reward new users immediately with useful contacts. Eventually, they will give even laggards the motivation they need to add their own pages. One of the main tasks for a knowledge champion, therefore, is to create enough motivation to maintain engagement in knowledge activities until they pass their tipping points.

For the staff at large this can be provided in many ways, ranging from publicity to a modicum of gentle coercion. Simple practical steps such as highlighting interesting project reviews in the company newsletter, encouraging contributions to the knowledge base by nominating a topic of the week, and insisting on procedural disciplines such as the completion of personal pages and project reviews can all be useful. More fundamentally:

- *Enthusiasm is infectious*, so it helps to have managers and champions who are enthusiastic about knowledge activities.
- *Desirable jobs are sought after* and unpopular ones avoided, so it is important for staff to see knowledge roles as desirable rather than chores. One way to do this is to give roles such as leading review workshops and moderating the knowledge base to the up-and-coming rather than to those who simply have time on their hands or can be best spared; better still, consider making experience in knowledge roles a prerequisite for promotion.
- *Recognition can be a strong motivator*, so it is good, for example, to include a list of contributors in knowledge base articles. If these are hot links to personal pages they serve knowledge purposes too: they help users to judge the trustworthiness of information, and they make it immediately clear who the helpful domain experts are, and easier to get in touch with them.
- One of the best *incentives for using knowledge tools* is that they make people's jobs easier or more rewarding, so they

should be designed to do that. It is not enough, for example, simply to provide information; it needs to be information that will help complete people's knowledge jigsaws and to be easy to find, understand and use. This means in turn that it needs to be written with the user's point of view in mind, well organised, well written, and concise.

- A *social component* makes activities more rewarding. Simply providing a section for personal interests in personal pages can give them social value, incidentally serving knowledge purposes by encouraging people to keep them up to date and become habitual users. Project reviews can have social value, too, if they begin or end with a sandwich lunch during which no formal business is done. In practice most people discuss work anyway.

- People like to know that their effort is worthwhile and that they are working for a successful organisation, so *quick wins* that show how knowledge management is working and helping the business are valuable. As Kotter puts it, they 'energise the change helpers, enlighten the pessimists, defuse the cynics and build momentum for the effort.'

- More direct use of *sticks and carrots* has a place, provided it is sensitive. The skills field in Broadway Malyan's *Who's Who* social networking tool, for example, shows the default entry 'I have no skills to offer' until this is overwritten, while Feilden Clegg Bradley are planning to make printouts of personal pages a prime reference in annual assessments, encouraging staff to record their experience and achievements promptly.

Subtle motivations like these have been found to be more effective than overt rewards such as prizes and bonuses. They reward everyone who makes an effort, and, though small, they can be frequent. Overt rewards, on the other hand, can easily backfire. They can demotivate those who do not receive them more than they motivate those who do, and habituation quickly sets in. They come to be expected, and then they lose what motivational power they had, and it becomes hard to phase them out without causing resentment.

Unfortunately, motivation can easily be negated by demotivating factors and obstacles, so it is important to identify and remove these. Managers can start the process by introspection, asking themselves why they do not do various things they are supposed to do. Many practices' QA manuals require every project to be reviewed, for example, but this rarely seems to happen. A staff survey can also be revealing. The barrier most often mentioned in survey responses is time pressure – which is ironic, given that saving time by reducing rework and the reinvention of wheels is one of the main aims of knowledge activities. When the pressure stems from specific conflicting demands such as billable time targets the solution is obvious: legitimise and create space for knowledge activities by giving everyone a time budget for them. Equivalent ways can be found to remove most other demotivating factors and obstacles.

Be obsessive about details

Details can make the difference between success and disappointment in many aspects of knowledge management. The design of procedures, software tools, workspaces, mentoring schemes and all the other enablers needs to be informed by a clear understanding of purpose, of the practicalities of day-to-day management and use, and of the viewpoint and psychology of users. It pays to think carefully about all these issues; involve users in design; pilot-test; ask testers questions about both overall impressions and detail; if possible, observe some tests; and be prepared to modify and retest until all the obstacles to effective use have been weeded out.

Different details matter in different contexts. In software systems the functionality needed to do what the user expects and the ergonomics of the user interface are always important (which is why companies like Microsoft invest so much in user testing). Social networking tools, for example, need to give users write access, provide search facilities tailored to their core purpose of connecting people with questions to colleagues who can answer them, and present the most useful material at the top of the page. In activities such as project reviews and mentoring, which rely on person-to-person interaction, management skills come to the fore. Interpersonal skills are crucial, so it is important to match personalities to roles. The higher (and more visible) costs of labour-intensive activities like these make resource allocation a consideration too: what proportion of projects should have a hindsight review, for example, how should they be chosen, and how long should reviews be? And there is a balance to be struck between insistence on what may appear to be the ideal approach, flexibility, and practicality; it may be necessary to design training as well as the procedures themselves. Failure to give adequate consideration to apparently secondary issues like these can compromise the whole effort.

Don't be too easily satisfied, but don't make the best the enemy of the good either

Learning and knowledge-sharing are not new. They have always gone on in professional practices, and indeed they must have entered into every collaborative activity since humans first started to communicate.

There can be few professional organisations today without a degree of knowledge awareness and conscious management of knowledge. It is meaningless, therefore, to ask whether an organisation has or has not 'got' knowledge management. The only worthwhile question is 'How well does it work?', and the answer does not lie in software systems and procedures but in what people actually do and how much value the business gets from it – in achievement, not appearances.

I know of companies that claim excellence in knowledge management on the strength of steps such as creating a knowledge manager post, setting up some intranet discussion forums, and writing a requirement to hold post-project reviews into the company QA manual, but whose actual achievement is minimal. No board should be satisfied with that. On the other hand, I also know of organisations that are so keen to make a real success of knowledge management that they are reluctant to take simple steps that would be immediately beneficial but which fall short of their high aspirations. That is equally misguided.

The best course is a middle one, aiming to achieve progressive improvement through a succession of manageable steps and a clear focus on practical achievement.

Have patience

The need for patience has already been mentioned several times in this chapter, but it is so important that it bears one final repetition.

It takes time to put systems in place, for desirable new habits to develop, and for benefits to show through. Even something as user-friendly and self-evidently useful as BP's widely admired *Connect* social networking tool needed patient nurturing: it took a year for the first 10% of staff to create personal pages and another three years for the next 20% to follow, despite knowledge management's high profile in the company. Professional services organisations should do better than this with social software, because a higher proportion of their staff are knowledge workers with PCs in front of them most of the time, but even so BP's experience is salutary. I know of no organisations that have made knowledge fully part of their mindset and practice in less than three years. Expectations need to be realistic from the start, and patience is indispensable.

Other roles

The purpose of knowledge management is to enable, facilitate and foster effective learning and knowledge-sharing as pervasive and everyday habits *throughout* an organisation. Ideally, *everybody* should take every opportunity that arises to learn, and share all their knowledge freely with everybody else who can make use of it. Learning and knowledge-sharing should be (and be recognised by management to be) a normal part of everybody's work, and as such they are activities 'staffed' by the whole organisation. Nevertheless, there are several identifiable roles in between top-level leadership and these everyday knowledge activities.

As we have seen, creating a vision and strategy, and sustaining them, are unavoidably tasks for the chief executive and other members of the company's top management team. None of this can be delegated effectively, although help from an outside consultant can be invaluable. But the work needed to realise the vision and implement the strategy – making the culture supportive, providing motivation within business units and project teams, developing software tools and knowledge resources, organising group knowledge activities, and removing obstacles – is largely a task for others. Some aspects clearly devolve onto middle managers, but who should take on the rest – the nitty-gritty of making knowledge management work? What skills do they call for, and how onerous are they? Are they also jobs for the middle managers, or are they part-time tasks for practitioners, or add-ons for existing support functions? When is it worth employing full-time knowledge management specialists?

Knowledge management has developed some of the characteristics of a minor profession over the past few years. It has a considerable literature of its own, a growing number of universities are offering MSc courses in knowledge management, and more and more companies are creating posts with titles such 'Knowledge Manager', 'Chief Knowledge Officer', 'KM Administrator' or even 'Director of Knowledge'. Some of this is genuinely new activity, reflecting a new awareness of the business value of knowledge and of the potential for creating more and using it better, but some of it is little more than rebranding of traditional information management and librarianship, somewhat expanded in scope. It is not clear whether knowledge management as a profession (or even as an academic subject) is a passing phase or is here to stay.

Whatever happens in this regard, I am convinced that there should be a board member in every professional services organisation with a specific responsibility for ensuring that knowledge is taken into account in all board discussions, for developing the organisation's intellectual capital, and for overseeing knowledge management. Knowledge is easily forgotten or sidelined, and it needs somebody to fight its corner. It is only appropriate, though, to create a *separate* board-level post in organisations that also have other functionally specialist directors.

The case for specialist knowledge *manager* posts is more contingent. Only organisations in which knowledge systems and practices are complex and demand continual development to keep up with changing business requirements can justify employing full-time specialists in knowledge management per se. Even then, it is difficult to make the posts attractive: too few are needed – only the largest organisations can justify more than one – to create a career structure, and the incumbents can easily become intellectually isolated. Many of the people who have acquired a high level of specialist expertise developing knowledge management in large organisations have left to form consultancy practices of their own. Narrowly specialist knowledge management posts are rarely an attractive option in professional services.

There is a stronger case for one or more 'knowledge managers' who combine system oversight, development, staff training in knowledge activities such as mentoring and leading hindsight reviews, and support to communities of practice, with more traditional librarianship or HR roles. The balance between these roles can vary. In the construction industry the most valuable knowledge tends to come from project experience, so Bovis Lend Lease employs full-time facilitators in several offices to put knowledge seekers in touch with experts elsewhere in the company, and record the more interesting questions and answers for future reference. Lawyers face the different challenge of a continuing flood of legislation and case law, and some employ 'professional support lawyers' to keep their client-facing staff up to date with this, analyse its implications, identify new business opportunities, and support the practice in other ways. WSP employs a group technical coordinator to write technical guidance notes, provide a help desk service, organise technical seminars, and maintain an electronic reference library and networking directory. Generally, the case for posts like these increases with the size of the organisation and the volume and business-criticality of the documentary information that needs to be managed. There is, though, no simple test for deciding whether full-time posts are appropriate and where they fit into organisational structures, or who should do the work when it is divided between several people. It is up to individual organisations to work out what will best suit their needs and circumstances.

The roles, fortunately, are clearer than the way they should be staffed. There are two phases to most knowledge initiatives. In the first, the vision is translated into a plan of action; knowledge processes, tools and a nucleus of content are developed and tested; and then they are rolled out. This evolves into a second phase, in which tools and processes become part of normal practice. During this phase and beyond, they need continuing oversight, development, and periodic review to ensure that they succeed in their purpose of supporting the organisation's business. In parallel, there is an ongoing stream of knowledge-handling tasks to be carried out, including acquiring knowledge from external sources, facilitating reviews, documenting lessons learned, collating and interpreting information, moderating forums and knowledge bases, training staff in knowledge activities, and supporting CoPs. Finally, knowledge needs to be taken into account in other aspects of management, including professional leadership, operations, marketing, project management, information management, IT and HR.

Providing the tools

Only a few years ago the first step many organisations took towards knowledge management was to look for an all-in-one software package to buy. That seems naive today. IT tools still have a large part to play, but a simple procurement process has been replaced by the more demanding tasks of selecting software capable of supporting several knowledge activities, integrating it with existing systems so that information can be shared between them, and developing logical frameworks for organising content, page templates, user guidelines and other ancillaries. In addition, guidelines are required for activities such as foresight and hindsight reviews, mentoring and communities of practice. All this calls for a combination of knowledge management expertise, familiarity with the organisation's work and, in the case of the IT tools, technical skills and sometimes graphic design skills as well.

The knowledge management expertise can be provided by in-house staff – either by sending someone on a specialist MSc course, or by recruiting an appropriately qualified specialist – but it is often best provided by an external consultant, for the reasons we have already discussed. Long-serving staff who have had specialist training have the advantage of in-depth knowledge about the organisation, but it is important to recognise that they will remain novices in knowledge management for some time. Consultants and newly-recruited knowledge specialists can be immediately effective provided they work closely with managers and users in developing strategies, tactics, tools and processes to meet the organisation's needs. Specifying and guiding the development of new software systems calls for close collaboration with IT staff as well.

There are numerous detailed choices to be made. What proportion of projects should have foresight or hindsight reviews, and how should they be chosen? Who should facilitate reviews, and what training do they need? How can mentoring best be used? How much autonomy should communities of practice have, and how much corporate support? Should they be left to arise spontaneously, or be set up to reflect corporate concerns? What sections should personal pages contain, and in what order? How important is WYSIWYG editing? Is it worth listing professional qualifications on personal pages, or is staff grade enough? Should all the projects that people have worked on be listed, or only recent ones? Should there be facilities for including photographs? Should personal interests appear high up on the page or near the bottom? In a knowledge base, what topics should be included, and how should they be organised – alphabetically perhaps, or in a hierarchy that logically reflects the way the users think and work? Should project examples be included in the knowledge base, or cross-referenced from a separate project directory?

The choices that are right for each organisation depend on fine details of the work it does and the kind of information users would like to have, and the users themselves are the best judges of that. Their perspective can be provided by small working groups, by a consultation process carried out by the knowledge expert, user surveys during the testing phase, or in other ways.

Specialist IT skills are normally best provided by in-house staff because choosing, installing and integrating appropriate software requires an intimate knowledge of existing systems: it takes time and costs money for external consultants to acquire this, unless they have done previous work for the organisation. However, the level of skill and effort required varies widely according to the software chosen and the degree of customisation and integration required to provide the necessary functional capabilities. The knowledge base at architects Feilden Clegg Bradley, for example, was installed in a few days by a junior architect with an interest in IT, using an open source wiki package as the basis. In contrast, Aedas, a practice several times larger and with more complex existing systems, devoted some man-months of professional programmer effort to developing software from scratch in-house. Which approach offers best value depends on the context, and an informed choice requires negotiation between the knowledge management and IT experts.

When tools have been developed, seeded where necessary with initial content, and rolled out, a close eye needs to be kept on them until they bed down. Teething troubles can turn users away, and it is often hard to attract them back, while underuse may indicate luke-warm support or even hostility from local managers, and call for swift action from the top. Beyond that, specialist expertise is required only to deal with any IT problems that arise, to audit the tools periodically, and to help in their further development; responsibility for continuing oversight can pass to the knowledge director.

Developing knowledge content

It is essential to seed IT systems such as social networking tools and knowledge bases with a nucleus of content before launching them for general use. Content is the bait on the hook, rewarding early users and showing them by example how to follow up with contributions of their own. Networking tools can often be populated with enough information from existing HR databases to make them useful imme-diately, if only as internal phone directories. But creating content for other systems can be challenging, because (unlike tool development) it can be done only by busy professional staff, who are the only people with the necessary domain knowledge. Knowledge specialists can only help to guide the process, remove obstacles, and encourage contributions.

Personal pages can be completed only by their owners. This is a short and undemanding task, but nevertheless one that disappointingly few people carry out without encouragement or even gentle coercion. Most plead lack of time – the face-saving rationalisation of a self-interested calculation that completing a personal page is likely to be rewarded less than getting on with the current project. Managers need to convince staff that the opposite is true. They need to take time to explain what the system is for, and how individual staff and the practice will benefit from it, and they need match their subsequent behaviour to the rhetoric. If they skimp, or give out conflicting signals – announcing the launch in an email instead of face to face, or showing impatience with staff who take time out of their project to complete their pages – it can take a long time to reach the tipping point where the system becomes such a rewarding source of information that using it becomes habitual, and keeping content up to date comes to seem natural. Even with good management, progress is often slow, but there are ways to speed it up. One of the best is to make completing a personal page part of the organisation's induction process: networking tools are particularly helpful to new recruits, and their pages help introduce them to their new colleagues. Involving joiners in knowledge activities as soon as they start sets them on the right road, and reinforces the development of a pervasive knowledge culture.

Completing an entry in a project directory is more time consuming, and whereas it is clearly cost-effective for everyone to complete their personal pages, the value of project information falls off sharply after a few years. This suggests a selective approach, with entries mandatory for new projects, encouraged for active and recently completed projects, and left to personal discretion for older ones. The strong ownership that project leaders typically feel for their projects makes them the obvious authors in the first place, but they should nevertheless not be allowed a monopoly; team members, hindsight review participants and others may all have valuable contributions to make. And, even if entries are made mandatory, managers have a vital role to play in spiking the lack-of-time excuse.

Developing content for knowledge bases is also time consuming, and it is intellectually more demanding. It is not self-evident what to include (as it is in personal or project pages), and it takes logical thinking and authorial skill as well as domain knowledge to select appropriate information and present it in a form suitable for just-in-time use.

The two most fruitful sources of initial content are existing guidance documents and personal knowledge. Guidance is rarely suitable to be quoted verbatim, but key ideas can often be identified reasonably easily and used as the basis for knowledge base entries. Links can be provided to the complete documents for the minority of users who want to probe further. Guidance can be adapted in this way by any experienced professional or a knowledge manager with a professional background; it can even be worth hiring a technical author, who may cost less. In the absence of existing documents to quarry, entries need to be solicited from acknowledged topic experts who can provide credibility as well as expertise, or from communities of practice where they exist. It is unrealistic to expect experts, who by definition are likely to be among the busiest people in the practice, to make substantial contributions on top of their normal work. This is one of the areas where top management should consider making an explicit investment by creating a budget for them to use and reducing their project workload temporarily in order to get the knowledge base up and running. Communities of practice have the advantage that they can share the work, while investing contributions with comparable authority and developing the next generation of expert contributors.

As soon as enough content is in place to make the system useful on selected topics it is timely to encourage contributions from other staff. They need to understand as quickly as possible that everyone is qualified to contribute, that contributions need often be no longer than a sentence or a paragraph to be useful and take a few minutes to make, and that their efforts are valued.

It is reasonably easy for people to see opportunities to contribute when they are involved in activities such as project reviews and communities of practice that are specifically designed to generate lessons and insights. It is more difficult to recognise when information and ideas from everyday practice are worth recording. To encourage people to develop the skill, it is helpful to make contributions mandatory in some circumstances. One obvious occasion is when people attend external conferences or seminars, and a knowledge base contribution can be made a quid pro quo for the practice's support.

Managers need to be consistent in supporting the development of content, and in ensuring that staff have time to make contributions and that other inhibitions are minimised. Knowledge specialists also have a role to play, helping contributors to understand how the knowledge base is intended to support the practice's work, how to recognise appropriate material, and – perhaps most important of all – how to present it so that users will find it useful in the rush of everyday work. The need for clear, concise presentation is easily forgotten, but it is vital if the effort is to deliver real value to individual users and to the organisation.

Ongoing knowledge activities

When initial development and testing have been completed, and tools launched successfully into use, most of the work involved in ongoing learning and knowledge-sharing falls to managers and staff in general. Ultimately, it should become simply a normal part of their job. There are, though, several roles that that need to be given to named staff. These include:

- *Knowledge champions*, charged with carrying the torch for knowledge in individual offices and professional specialisms, liaising with the knowledge managers and knowledge director. They need to ensure that the knowledge perspective is heard in management, that new staff are introduced to knowledge activities, successes publicised, and so on.
- *Workshop facilitators*, to provide skilled leadership for foresight and hindsight events and make them as productive as possible. Facilitators may also have part of the responsibility for recording lessons learned on the knowledge base, or that may be a separate role.
- *Knowledge base moderators*, responsible for keeping an eye on contributions to their designated topics to ensure that contributors follow the rules and have adequate writing skills, maintain the quality of content, and edit where necessary.
- *CoP chairmen*, to provide professional leadership and initiate CoP activities.
- *Librarianship*: managing the acquisition of information from external sources and making it part of the organisation's knowledge resources by means such as adding links to it from the knowledge base, bringing topical material to the attention of appropriate people, and analysing the implications for the organisation of forthcoming legislation, new standards and so on. Specialist librarians may not be needed; tasks like these can often be shared between various people, but everybody does need to know where the responsibility lies.
- *Periodic audit* of knowledge tools, processes and activities, their strengths and weaknesses, and the business value they generate, to ensure that they are fulfilling their purpose and to suggest improvements. Ideally, this should be done by someone such as an external consultant who is both visibly

independent of vested interests and in touch with developments in knowledge management.

Knowledge-conscious management

Most aspects of management have a knowledge dimension, quite apart from any overt knowledge activities in which managers may be involved. Professional services organisations are increasingly using an effective knowledge management system as part of their offering to clients, and client contact before, during and after projects can be an invaluable source of knowledge. An understanding of how people learn can inform CPD and career planning, and knowledge activities can be used to help their professional development and prepare them for new roles. Effective use of organisational knowledge can boost productivity and quality, and reduce risk. Social networking tools can help select the best qualified teams, and provide useful summaries of experience and skills to use in annual assessments and promotion reviews. Knowledge activities need to be included in budget planning. In short, to maximise the business value of knowledge it needs to become part of the corporate mindset and inform every aspect of management. Knowledge management is not a simple purchase, or an item to be ticked off on a board's action list and forgotten; it needs to be made a pervasive and permanent part of management and professional life.

The next chapter discusses the first practical step on the road to achieving that: carrying out a knowledge audit to find out how well basic knowledge processes, and any tools and techniques that are already in place, are working in an organisation, and putting the conclusions together with strategic priorities to develop an action plan.

Chapter Six
Knowledge Audit and Beyond

Finding square one

Strategy and vision are vital guides, but practical action needs to start from the realities of existing culture and practice. How well are we learning at the moment? How freely is knowledge flowing round? What's working well, and what isn't? What's getting in the way? We can imagine any future we want, but the present is a matter of unalterable, and often awkward, fact. When you embark on a knowledge initiative it is just as important to understand where you are starting from as it is to have clear aims and objectives. The process of finding square one is usually called *knowledge audit*.

Audits at this stage have two other purposes that are equally important. First, they can help to inspire conviction that change is needed. Unfocused ideas such as 'we really must look into knowledge management', or even 'knowledge is our greatest asset, we really must learn to use it better', are not strong enough motivators to drive the sustained effort involved in making a success of a knowledge initiative. Major change needs emotional engagement as well as cognitive judgements, and one of the best ways to achieve this is to confront specific examples and hard evidence. That is why graphic photos and personal stories are the mainstay of charity appeals, and it works for business initiatives, too. Second, audits can help shape the details of initiatives and set priorities. Business strategy defines the broad objectives for knowledge management; knowledge audit defines the tactics needed to achieve them.

When knowledge management is already an established part of 'how we do things round here', simpler audits carried out every year or two are invaluable for monitoring its health, revealing weaknesses, and showing how it can be improved further – just as financial audits are for financial management.

Audits can enquire into either or both of two different things:

- *Knowledge processes*: the IT tools, information services, techniques, procedures and habits through which knowledge is accumulated and shared in an organisation, how they are used, and how the organisational and cultural context influences their effectiveness
- *Knowledge assets*: the actual knowledge and information in the organisation – what there is, where it is, and what business value it has.

An audit of *processes* in a professional practice will typically review:

- The *formal mechanisms* – including libraries, online information services, databases, intranets, search tools, groupware, project review procedures, training and mentoring programmes, and the other tools and techniques discussed in this book – and how they are perceived, used and valued
- *Organisational factors*, such as geographical dispersion, management and work group structures, workspace design, time-booking systems, performance targets, staff appraisal metrics and reward systems, and how they influence knowledge activities
- *Cultural norms and values*, such as whether staff typically keep personal libraries, how much they talk to nearby colleagues, whether they would phone a colleague they have never met to seek advice, and their beliefs about management's attitude to knowledge activities.

An audit of *assets* might assess:

- What kinds of knowledge are most *critical to business success*
- What kinds and amounts of *recorded knowledge* are held in the various formal repositories such as libraries and databases
- What *tacit knowledge* (knowledge in people's heads) exists, where, and how accessible it is – who the key experts are, what their expertise is, how widely their expertise is known, and how well they share it with other people.

Other kinds of knowledge asset, such as brands, patents and registered designs – what is often called 'intellectual property' – are important in many other industries, but rarely in professional services.

Audits can provide answers to a wide range of questions, from the simply factual ('How many people contributed to the knowledge base last year?') and tactical ('Are the new personal pages helping staff to extend their networks?' or 'How can we get more value out of post-project reviews?') to the strategic ('Are we operating like an integrated practice yet, or are we still like a collection of independent offices?'). With such a wide range of possible questions, a comprehensive approach is impracticable. Even if the cost was acceptable (and that is unlikely), there is a more fundamental constraint: much of the most important data can only be obtained from staff, and the process draws on a limited pool of goodwill. Most people are happy to contribute to a process that promises to make their organisation (and them) more successful, provided it is professionally carried out and makes reasonable demands on their time, but they quickly lose patience with *un*reasonable demands or an inept process. Fortunately, the Pareto principle applies as much in this case as in so many others: 80% of the value comes from 20% of the effort. Knowledge audits (like financial audits) can – and should – be scaled and shaped to suit the organisation, focusing on the issues and processes that matter most to the practice, and the places where investigation promises to be most fruitful.

One thing a knowledge audit can hardly ever do is *quantify* the value of knowledge processes or assets, and hence the financial return on knowledge management activities. It is possible occasionally: Xerox, for example, was able to compare the efficiency of a group of copier service technicians who used their Eureka tips database with the efficiency of a group who didn't, and show that using Eureka cut costs by 10%. But work in professional services is too varied for approaches like this to work, and we have to fall back on indirect indicators – and remember always that they *are* only indicators, not measures, with all the limitations and uncertainty that implies.

Indicators need to be chosen and interpreted with care. More is not necessarily better: unused technical databases, skills directories full of out-of-date information and project 'reviews' that have degenerated into box-ticking ritual are costs, not assets. Activity is generally a better indicator than volume (busy people do not choose to do unproductive things), but the benefits of even a well-used knowledge base may not justify the cost of maintaining it. To add further difficulty, the benefits of knowledge management arise unpredictably and unevenly, and the exchange of one idea that leads to a brilliant innovation or avoids a major mistake can pay for years of knowledge management. It would be fortuitous indeed to observe an event like that during an audit. Even without hard metrics, though, it is always clear to management and staff when knowledge systems and activities are working: they know when they help them, they can see when they work for other people around them, and they have success stories to tell. Indicators like these are soft, but they reveal the truth. Quantitative measures can be more telling at the level of individual systems and processes, but the softer ones are often the best here, too. Productive hindsight reviews are lively, run over time, and participants leave with smiles on their faces; up-to-date personal pages show that networking tools are valued; useful knowledge bases grow.

When an organisation wants to start managing its knowledge systematically for the first time there are likely to be few purpose-designed tools and processes to assess. However, learning and knowledge-sharing go on everywhere, however unthinkingly, and there are always systems, activities and working practices to look at and, potentially, build on. In a context like this, an audit can assess such things as the extent of people's personal networks, what people do individually and in teams to learn from experience, what documentary knowledge resources exist, how accessible they are, and how much people use them.

However little or far systematic knowledge management has developed, topics that have only recently become important (as a result of changes in legislation, for example) make good litmus tests for learning and knowledge-sharing. Corporate capability in these depends much more on the effectiveness of internal knowledge systems than it does in longer-established areas. Sustainable design is one such topic in construction at the moment. People who qualified more than a few years ago are unlikely to have learned much about this at university, and although recent graduates will have been introduced to the principles, they are unlikely to know much about the practical details that make the difference between success and failure. At the time of writing (2007), only a few people in a typical design practice will have been personally involved with any low-energy buildings right through from design to commissioning and occupation. Topics like this provide telling indications of learning and the flow of knowledge.

The variety of scale and focus that is appropriate in audits makes it impossible to give step-by-step recipes. However, the basic data-gathering techniques of collecting factual information, interviews, questionnaires, activity observation and social network analysis are universal, applicable across most kinds of organisation, and potentially relevant in both process and asset audits. In practice, there is considerable overlap between the two: it makes more sense, for example, for interviews and questionnaires to investigate processes and assets together than to run separate surveys.

The basic sequence of collecting data, analysis, assessment, and reaching conclusions about future action is universal, too, although not necessarily linear. It is often better to proceed iteratively, collecting data from just a few people and perhaps just one part of an organisation in the first place, analysing this, and using the understanding that results as the basis for designing a second and more focused phase of data collection that draws on more people and sources. It is up to the auditor to choose which of the various techniques to use, how, and on what scale. Whatever follows, though, the first step has to be taking stock of the basic facts about systems and knowledge assets: it is impossible to ask detailed questions about a knowledge system or judge what activities are worth looking into without knowing at least in outline what processes or assets exist.

Audit techniques

Stocktaking

Stocktaking involves collating factual information about relevant databases, software tools, book holdings, journal subscriptions, training programmes, staff appraisal criteria, project assessment systems, knowledge-related elements in job descriptions and so on, together with any available records of related activity. In single-discipline professional services organisations, where directors are personally involved in all aspects of the business, this may be quite simple. They have a broad knowledge of what goes on, and interviews with two or three of them, and with selected middle managers and functional specialists, can often avoid the need to wade through voluminous documentation. Together with direct observation of key systems and processes (the interface design and functionality of software systems, for example), interviews make an efficient way of collating a picture good enough to show where staff surveys and other more detailed investigation should be focused. At the same time they can give insights into attitudes, frustrations and aspirations that are invaluable in planning a knowledge initiative. In organisations of up to 1000 or so staff, interviews and observation like this may be an adequate stocktake in themselves.

In organisations that are larger, that are multidisciplinary, or where management is more divided by function, it may be necessary to look at documentary material as well, and speak to more people, in order to get a sufficiently complete and unbiased picture. Useful material may include the following:

- Available *statements about company policy* (on training, for example).
- Statements about the *purpose* of individual processes and systems (often included in budget proposals)
- *Descriptions* of processes and systems from existing documents, interviews, or based on direct observation. User guides and illustrative material such as screenshots (a picture really can be worth 1000 words) and examples of database records and completed forms can be particularly helpful.

Specifications are less so; they are often too detailed to be useful in a knowledge audit, and may fail to cover the aspects that matter. When documents fail to give a clear picture, interviews or direct observation become indispensable.

- *Activity data* from sources such as project files, library loan records and database access logs. Many knowledge systems are more appearance than substance: QA manual requirements to hold post-project reviews that in fact are rarely carried out, intranet discussion forums that nobody visits, or guidance documents that are hardly ever consulted. Without evidence on usage, an audit can give a completely false picture. Fortunately, most formal knowledge activities create records of one kind or another. The frequency of processes such as post-project reviews will be reflected in the number of reports on file; forum usage is normally shown on-screen (and can anyway be discovered from system logs); knowledge base usage is likely to be detailed in IT system logs; and so on. System logs can often provide more than simple event counts, revealing who has used the system, and what information they were looking for (or were contributing). This is labour-intensive to analyse exhaustively, but even a small sample can yield useful insights.
- Other *quantitative data* such as numbers of journal subscriptions, records in databases, courses attended etc. This need only be approximate.
- *Geographic location* where appropriate, for example of libraries and material samples.

Stocktaking has a variety of uses beyond giving focus to the knowledge audit. The results can be valuable, for example, as a reference when designing improvements to knowledge systems, as a baseline for assessing their success, and as the basis of user guides both for people with knowledge management responsibilities and for staff in general. However, the process can be more time consuming than it appears at first sight, particularly in organisations with several semi-autonomous offices or business units. It is important to keep asking how much value the next piece of data will add, and to be prepared to compromise or stop. When effort is limited, it should be focused on the systems and processes that seem most likely to repay further investigation – always bearing in mind that appearances can be deceptive, and that priorities may change as understanding grows.

Interviews

Interviews really come into their own as a tool for investigating how well the reality of knowledge processes matches the intent. Between them, senior staff, IT specialists and librarians usually have a shrewd idea of which systems and processes are widely used, neglected, liked and disliked, as well as knowing which exist, and interviews with them can investigate both. However, senior and specialist staff can be surprisingly ignorant of other users' feelings: the design of systems and processes is often shaped predominantly by management needs, and the result can seem successful to managers while being thoroughly disliked by staff in general. The MIS in one practice that I audited, for example, asks users to update their time sheet every time they log on; the nag message produces excellent management data, but staff loathe it. Top management can also have an unduly rosy (or unduly cynical) view of behaviour. Someone who has designed a project assessment process may be reluctant to recognise that it is treated as a box-ticking exercise from which nobody ever learns anything. To avoid being misled about the realities of day-to-day practice it is important to interview staff of all grades.

A semi-structured approach based on an initial document-based stocktake is perhaps the most generally useful. This ensures that interviews cover all the main processes and issues while leaving the door open for new ones to enter the discussion, and reducing the potential for preconceptions to bias answers.

A skilled interviewer may be able to take adequate notes without significantly interrupting the conversational flow. Nevertheless, it is wise to record interviews as well, if only as a backstop.

Social scientists have developed sophisticated methods for analysing interview records, but these are highly labour-intensive. They are unnecessary in knowledge audits, especially when they are going to be followed up by a questionnaire. Interviewees are typically open and forthcoming, and what they say can be taken at face value, if only as an expression of personal experience or opinion. That is enough to provide the basis for designing a questionnaire.

Questionnaires

A knowledge audit can be thought of as a process of gradually bringing a fuzzy picture into focus. The auditor starts with a patchy, broadbrush and uncertain understanding, and adds firmer outlines and increasing detail as the evidence builds up. Interviews are ideal in the early stages, because they can start with almost no knowledge. Questions can be rephrased on the fly, and if a line of enquiry proves unproductive it can be abandoned with nothing lost; each interview can be made better than the last. This flexibility is particularly helpful with senior staff, who have more individual roles, are more likely to have thought about information and knowledge, and are usually the most articulate. However, conducting interviews and analysing the results is too labour-intensive to be done on a large scale. They show what is likely to repay further investigation, but an all-staff question-naire is a more cost-effective way to get representative data. The response rate to a well-designed questionnaire in a professional prac-tice can be expected to be 25–50%, enough for conclusions to be drawn with confidence about the practice as a whole and at least about the larger offices. The two techniques are complementary: interviews give rich insights that other tools cannot match, while ques-tionnaires (together with direct observation and social network analy-sis) are indispensable for collecting robust evidence to test and support them.

Questionnaires are the Swiss Army knives of knowledge audit. They can give invaluable insights into almost all aspects of an organisation's knowledge management – assets as well as processes. However, with limited goodwill to draw on, they have to be used sparingly, and designing a questionnaire capable of yielding reliable and operation-ally useful information is not a trivial task. It is a subject in itself (Amazon lists several hundred books on 'survey methods'), and the process takes skill, domain knowledge, and almost obsessive care. Most of us have been irritated by professionally designed surveys that include ambiguous questions, demand information we do not have, or leave us with the nagging feeling that we have been prevented from expressing our real opinions.

Common pitfalls include:

- Making questionnaires too long. The limit of most people's patience is about 15 minutes, which typically translates to about 60 questions – fewer if answers need more than brief consideration.
- Using the wrong kind of question. Both multiple-choice and open-ended questions (which respectively offer a short list of possible answers for respondents to pick from, and allow respondents to write in whatever they like) have their place, but are often used inappropriately. Multiple-choice questions yield data that is easy to analyse and likely to be statistically robust, but they take some skill to frame well. Open-ended questions are easier to frame, but labour-intensive to analyse, and rarely give statistically robust data.
- Failing to anticipate all possible answers to a multiple-choice question. There should be an answer to suit everyone, if only 'other' or 'not applicable'. It annoys people to be unable to give the answer they want, and that both wastes a question and makes people less inclined to take care with following ones.
- Asking questions that are too vague. It is worth asking a few general questions such as 'How useful do you find the . . .'

(overall impressions are important), but more specific and factual ones, like 'How often do you use . . .', with response options such as 'several times a day', 'about once a day' and so on, are better at pinpointing strengths and weaknesses and showing how tools and processes could be improved.

- Combining questions to which respondents may want to give different answers, as in 'How often do you use the new people pages and knowledge base?'
- Reading too much into replies to questions that ask respondents to speculate about their behaviour in hypothetical situations. Questions like this can be useful indicators of attitude, but responses cannot be taken at face value: people are usually poor at predicting their future behaviour (the 'knowing–doing gap' again).
- Asking for a judgement that respondents are not qualified to make, such as the survey thrust into my hand at a music festival that asked 'Do you think this festival is important in establishing Mytown as a major player in the British cultural scene?' Without researching what competing festivals there are, and how many people they attract, how should I know? (And that from a company supposedly expert in survey techniques!)
- Using words sloppily, such as 'regularly' instead of 'frequently'. Christmas comes regularly, but not frequently.

Poorly designed surveys are, at best, missed opportunities; at worst, they can be thoroughly misleading. Ideally, therefore, questionnaires, like interviews, should be left to people who understand survey methods, as well as the processes being audited and their business context. External knowledge management consultants can help here. Their independence helps too, because it encourages franker responses.

Social network analysis

Social network analysis (SNA) is a formal technique that can give unparalleled insight into where people turn for information and advice, and how tacit knowledge actually flows between people, groups and offices – and can reveal where they do *not* turn, and where knowledge is *failing* to flow. The results can be invaluable in auditing both systems and assets.

SNA uses carefully designed questionnaires to collect raw data, and specialised (often computer-based) tools to analyse it and reveal patters of interaction.

In the artificial but realistic example in Figure 6.1, it is clear even at a glance that contact between group A and group B is poor, and that it relies heavily on one or two people.

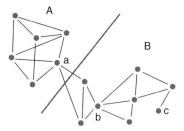

Figure 6.1 A social network map.

In this case, most people are reasonably well connected. However, communication between the two groups depends entirely on two individuals, a and b, and if *either* of them were absent it would be in danger of ceasing. All communications with group A pass through **a** (a 'gatekeeper' in the jargon of SNA). Group B operates almost as two subgroups, connected only by their common member **b**. Person b is well connected in both, and in frequent contact with five colleagues (both a 'boundary spanner' and 'network builder'), in stark contrast with c, who has only one regular contact.

The implications depend on the two groups' roles, and on people's individual expertise. If the groups do very different work the weak links between them may not matter, but if they work in similar fields it could be an important barrier to knowledge-sharing. Person b may be richly connected because he or she is inexperienced, and seeks a lot of advice, or an expert overloaded with requests for help. Some SNA diagrams use arrows to show the predominant directions of knowledge flow.

Results like this can show where action to improve person-to-person communication would be particularly helpful, and can suggest what form it should take. Depending on circumstances, this might be encouraging an isolated expert to be more helpful, recognising the contribution made by someone who is overloaded with requests for help and relieving him or her of some other work, putting people from two non-communicating groups together in a project team, or organising joint events where people can get to know each other.

Activity observation
Some aspects of behaviour are outside the reach of surveys. People are good at making simple qualitative judgements about their own experience – about whether they find systems helpful or easy to use, for example – but they cannot be expected to be analytic about things outside their area of expertise, such as how well project review workshops are led. Nor are they likely to be accurate in quantifying unconscious behaviour, such as how often they talk to passing colleagues

(and a surprising proportion of everyday actions *are* unconscious). Reliable evidence about activities like this can only be obtained by suitably qualified observers.

Activity observation is invaluable for understanding the reality, as opposed to the intent, of a variety of knowledge-critical behaviours. Examples include the use of IT system interfaces, the use of office space – desk surfaces, document storage facilities, display boards, breakout spaces, tea rooms and so on – and interaction with colleagues, whether it is casual (someone passing the desk) or structured (such as in a project review workshop or an appraisal interview). Companies producing mass-market software carry out extensive usability testing on interfaces using sophisticated techniques such as eye-movement tracking and key logging in order to find out how ordinary users behave, and to identify which facilities they use, which they fail to notice, and which they have difficulty with. It would be impracticable to go to these lengths with a knowledge base, but a short period of careful observation can nevertheless pay dividends. Equally, the intent at a project review workshop may be to gather input from all the members of the team, explore the causes of problems, and learn lessons, but only independent observation can reveal reliably whether workshop leaders are skilled in quietening the voluble, encouraging the shy, ensuring that sensitive issues are not evaded, and tracing consequences accurately back to causes.

From audit to action plan

A good audit gives rich insights into the strengths and weaknesses of existing knowledge practices, and provides a sound foundation for planning improvements. When combined with a clear view of business objectives, the results can help in deciding what additional systems and procedures to develop (if any), highlight issues that should be taken into account in their design, provide the basis for an action plan, and set a baseline for measuring progress. It is counterproductive to try to make progress on too many fronts at once, and choices need to be made.

There are too many factors to take into account in planning to discuss them all here, but a brief review of the most important will serve to illustrate the kind of issues that arise: business strategy; tactical objectives; strengths, weaknesses and opportunities; and constraints.

Business strategy
As we saw in Chapter 3, different strategic ambitions set different priorities for knowledge management. Raising quality and moving up in the market (a product leadership strategy, in Treacy and Wiersema's

terminology) puts the emphasis on developing high-level professional expertise and capability, and on people-centred techniques such as mentoring, foresight and communities of practice. Raising profitability, on the other hand, calls for operational excellence, and puts more emphasis on giving people quick access to the information they need in routine activities, and so on tools such as knowledge bases and project directories. Staff in design practices typically estimate that they spend around 15–20% of their time reinventing wheels and doing rework, and a similar amount looking for information: even a modest reduction in this can make a major difference to profitability in a business where staff are overwhelmingly the largest cost.

Tactical objectives

Detailed priorities are shaped as much by short-term pressures on the business as by broad strategy. The imminent retirement of senior staff is one common short-term pressure that calls for a tailored response. Another is a booming market, and the rapid growth in staff numbers and high levels of staff turnover that it leads to. Failure to support new staff makes them slower to become productive, and an influx of newcomers can lead to an insidious decline in standards and dilution of a practice's ethos unless steps are taken to integrate them quickly. Appropriate knowledge management can mitigate the effects of all these problems. Giving departing seniors time to mentor their successors and pass on some of their wisdom, and giving new entrants near-peer mentors to whom they can turn without embarrassment for introductions to colleagues and basic information about office procedures can make a real difference. So can providing software tools to help joiners to develop personal networks, and well-structured and well-stocked knowledge bases to give them the technical information they need for quality work.

Strengths, weaknesses and opportunities

Strengths, weaknesses and opportunities revealed by the audit are a further influence on priorities, and the main influence on the details of implementation. The commonest motivation for an audit is dissatisfaction with the status quo, so weaknesses have an unfortunate tendency to predominate. Architects, for example, have no tradition of looking back at completed projects, so practices such as hindsight reviews are rare. Networks are often surprisingly small, seriously restricting informal knowledge-sharing. Even basic information about projects is often hard to find, to the frustration of people preparing bids against short deadlines. Where there are communities of practice, some are likely to be vibrant while others are moribund. The implications for knowledge management of weaknesses like these are usually fairly obvious.

Strengths, and the opportunities they create, may be less so. These often arise from local initiatives, and come clearly to light only when analysis of questionnaire results shows marked differences between offices or professional disciplines. Successful local initiatives can be invaluable. They can get new practice-wide tools and procedures off to a flying start, and provide:

- Detailed examples to learn from (and improve on)
- Convinced early adopters with success stories to tell
- A nucleus of procedural guidelines, software or content that can be exploited immediately as well.

People who show enthusiasm and understand the basic concepts of knowledge management, and local caches of documented knowledge, can be a strength in themselves. Finally, there may be opportunities that arise from other local circumstances and from reusable information available in existing systems such as HR databases. Particular expertise, a strongly felt need that knowledge tools can help to meet, or simply available effort, all offer opportunities to make progress more quickly or with greater prospects of success.

Constraints

The most serious of these is usually a reluctance to back expressed intent with the staff time needed to realise it in practice. We have discussed this elsewhere as a challenge for business leaders; it is a challenge for knowledge programme planners as well. At one level it is, of course, simply one example of the knowing–doing gap so eloquently analysed by Jeffrey Pfeffer and Robert Sutton, but it tends to be exaggerated by other factors in professional practices. This is not the place to discuss them; it is enough to say that, whatever the priorities suggested by strategy and audit findings, it can be better to begin with the tools and procedures that make fewest demands on the time of the senior and middle managers, and to start developments that are unavoidably labour-intensive on a small scale. Fortunately, time from IT staff is usually more readily available.

A recent knowledge audit in a rapidly expanding and very busy practice led to an action programme designed to accommodate severe constraints on staff time. It started with four components:

- A new *networking tool*, to help new entrants to find their feet more quickly and staff in general to develop wider networks. Social tools like this demand a considerable amount of staff time in aggregate, but they make concentrated demands only on the time of IT staff. Managers only need help design the structure and interface and introduce them to staff, and the effort of creating new content is thinly spread across the whole workforce, typically calling for only 15–30 minutes from each member of staff. This is small enough to be absorbed into the normal working day, especially as it can be spread across several sessions. For new joiners, completing a personal page can be made part of the induction procedure, and done before they are expected to become productive.

- *Work-in-progress displays*, to help people to keep in touch with what their colleagues were doing, and to stimulate discussion and networking.

- Selective *post-project reviews* to extract more lessons from experience about design and process. These can be difficult to arrange, but participants find them rewarding and enjoyable, and the time investment can be made piecemeal – each review brings its own return. Again, the demands on staff time are spread widely enough to be absorbed without explicit budget provision.

- A *common software framework* to support both a knowledge base and a design-oriented project directory. The development of content is the chief stumbling block in systems like this. To get over it, the plan called for initial content to focus on just one technical topic where there was an opportunity to exploit an existing (and previously underused) resource, and on new projects. The effort involved in completing a project page was considered acceptable for project leaders, who would both have the information easily to hand and be among the main beneficiaries of the directory. It was also a consideration that the IT manager was an enthusiast for open-source software, and the basic software for both these two systems and the networking tool would cost nothing.

Other combinations of audit findings and constraints would, of course, point to different choices.

Putting plans into practice

In the next part of the book we turn to the main techniques and software tools that have proved to be useful in professional practices: designing offices to encourage and facilitate informal knowledge-sharing (Chapter 7); social networking tools to help people develop a wider range of contacts, and to locate specialist expertise (Chapter 8); mentoring (Chapter 9); processes to enable more to be learned from everyday practice (Chapter 10); communities of practice (Chapter 11); techniques and software tools for recording the new knowledge that arises from everyday practice, and developing an organisational memory (Chapter 12); personal knowledge management (Chapter 13); and the synergies that multiply the value of individual tools and techniques when they are designed to work together (Chapter 14).

Part Three presents case studies that describe how over a dozen professional practices and related businesses have grappled with the challenge of managing knowledge. Most were knowledge management virgins at the time. They had habits, procedures and IT systems that more or less incidentally helped to share knowledge (though rarely any to help learn from experience), but few had thought systematically about knowledge as something that could, and should, be consciously managed, let alone had any strategy for doing so. Their experiences illustrate some of the difficulties they found, and show how some of the tools and procedures discussed in Part Two have been put into practice.

Hints and tips

The idiosyncratic business strategies, needs, priorities and cultures of different organisations make the detailed implementation of knowledge management, and the paths taken by knowledge initiatives, vary widely. Each has to find its own path, and detailed guidance is impossible. Nevertheless, there are several hints and tips that experience has shown to be universally useful:

- *Avoid trying do too much at once.* Attention and effort are always in short supply for non-essential activities in busy organisations, and it is better to concentrate what there is on a small number of tools or techniques at a time. Social networking tools and hindsight reviews make good starting points because they are easy to understand, require relatively little time investment to make them useful, and help make the point that knowledge management is everybody's job. Besides, people enjoy well-run hindsight reviews.
- *Search out and support early adopters in key positions.* We have discussed this before, but it is so important that it bears repeating here. People are strongly motivated only by their own interests and needs, and the corporate case for knowledge management only enthuses business leaders. Other people will only give it enough attention and time to make it work when they see how it helps them with their more parochial interests and problems. Finding people who do see this is one of the keys to launching knowledge activities successfully: they are the natural early adopters, and their

Continued

enthusiasm is indispensable for starting the ball rolling. But enthusiasm leads nowhere without time to pursue it, so the early adopters need to include some of the managers who set budgets and deadlines. They are key; without them, even the best-planned initiatives wither.

- *Sideline opposition.* It will be disarmed when visible benefits start to flow.
- *Aim to give users more than you expect from them at every stage.* Make sure, for example, that there is enough content in knowledge bases to make them useful before you expect people to contribute material themselves.
- *Make a reality of engaging grass-roots users in the design and testing of software tools* to ensure that they meet *their* needs, not just what managers imagine to be their needs.
- *Debug IT systems thoroughly before rolling things out on a large scale.* Even minor bugs can turn people off, and it can be hard to get them to try again.
- *Resist the temptation to react immediately to suggestions for 'improvements'*, especially when they come from people who are not typical users. Let individual comments accumulate, take them into account alongside results from user surveys, and follow the consensus view.
- *Be realistic: don't expect everyone to engage actively, or benefits to flow steadily.* Experience shows that most of the work on knowledge bases and communities of practice is always done by an enthusiastic minority. The majority are simply users or non-contributing onlookers; Wikipedia is an extreme example, with 60% of content contributed by just 2% of users. That does not matter. All knowledge-sharing is a kind of market, and markets need consumers just as much as they need producers. However, the majority should be given continuing encouragement to become contributors, if only occasionally. Benefits follow the same pattern, with much of the value coming from a minority of cases. That has to be accepted, too.
- *Remember that only strong and sustained drive from the top can create a learning organisation, and only grass-roots effort can make a continuing success of it.*

Part Two
Tools and Techniques

Chapter Seven
The Knowledge-Friendly Office

Environments matter

We all know that our physical surroundings can affect our mood, and people have suspected for at least 80 years that they affect our work, too. But which aspects of the environment matter, and how? Research into the effect of workplace design on performance has given mixed signals. Studies at a Western Electric plant in Chicago in the 1920s suggested that the environment had little effect; Frederick Hertzberg's studies in the 1960s that it was a 'hygiene factor', which could degrade performance but not improve it; more recent work that a good environment might raise productivity by as much as 50%. A report by the UK Commission on Architecture and the Built Environment in 2005 suggested that workplace design can affect performance by 5% for individuals and 11% for teams. Office design has varied with the prevailing theories of work. The regimented compartments of the 1950s (too isolating) gave way to 1960s Bürolandschaft (too public), 1980s 'universal planning' brought back identikit cells, and in the 1990s variety returned in 'alternative officing' and Frank Duffy's 'dens', 'clubs', 'hives' and 'cells'. For some people the 21st century office has dissolved into cyberspace, a laptop and a Blackberry. Today, there is a consensus that design matters, but the details of cause and effect remain unclear. One of the few areas of reasonable certainty is that the physical arrangement of workplaces determines how people move around in them, and that in turn has a profound effect on the casual interaction and knowledge-sharing at the heart of a learning organisation.

With hindsight, the broader uncertainty about the effect of workplace design is not surprising. Work is influenced by many interacting factors, of which the environment is only one. Theories that focus on one and ignore the others apply only in limited circumstances (which are rarely spelled out), and there is no overarching system theory that links them all. In the Western Electric studies, performance gains apparently associated with environmental improvements famously turned out to be the result of management attention, just as attention from a doctor has been found to improve patients' condition without any actual therapy. The context-specificity of research results has often been forgotten: what works in a factory or a call centre is unlikely to be equally good in a professional office. And the nature of work, the tools we use, the structure and culture of organisations, and personal expectations have been transformed since the 1920s, especially in the past 20 or so years, changing the whole system almost beyond recognition. The emergence of knowledge as a central factor in organisational performance has been one of the most profound changes of all.

It used to be axiomatic that knowledge flowed from the top down, and that interpersonal interaction in the workplace should be strictly controlled, channelled through formal meetings and letters. Time spent chatting to a colleague was seen as time wasted, and email seemed dangerously uncontrollable. As recently as the 1980s incoming mail in some companies was all delivered to directors to redistribute, so that they could keep an eye on everything. Vestiges of those attitudes survive in a few places, but they have been largely swept away by competitive pressure and the flattening of hierarchies. At the same time, research on organisational learning has shown that knowledge-sharing between peers and near-peers is at least as important as the flow from the top down, and that far from being a waste of time, informal interaction is often the most effective and efficient way to achieve it. One-to-one conversation is a uniquely powerful mechanism for many kinds of knowledge transfer, as we saw in Chapter 2. This pattern is particularly evident in professional services organisations, where most people share a broadly similar educational background and do broadly similar work. In this situation, top-down knowledge flow becomes the exception rather than the rule. Formal communication dwindles in importance, too. Almost inevitably, it tends to focus on big and simple issues, and to sideline the subtleties and complexities that are crucial to success in most professional activities; in many organisations it is more likely to be ignored than respected. The ad hoc, timely and highly focused chat that resolves a discrepancy, clarifies an ambiguity or provides a key piece of information quickly when need arises is much more valuable to practitioners – and information gained in a casual conversation round the water cooler or a remark overheard can be equally so.

There is widespread agreement today that the design of workplaces has a major influence on this informal and casual communication. It is common experience that different offices can have radically different atmospheres, even in the same industry: some are static and quiet, others busy and buzzing. In knowledge-intensive professional work it is self-evident that differences like these must have consequences. To succeed today, management theorists suggest, organisations need to be 'ambidextrous', able both to exploit their existing business and to explore new directions. They need to 'surf the edge of chaos' – the narrow margin between the well-regulated order that makes for efficiency at doing familiar things and the anarchic freedom that responds most creatively to new challenges. The best workplaces are ambidextrous too, surfing the edge between the privacy needed for concentrated, individual work and the sociability that maximises informal interaction and knowledge-sharing.

From a knowledge perspective, therefore, a good workplace appears to be one that facilitates and judiciously encourages all kinds of informal (but intentional) and serendipitous (unplanned) interaction, from seeing what colleagues are working on while walking through the office, through conversations round the coffee machine, to ad hoc team meetings for brainstorming or reviewing design ideas – while at the same time lifting spirits and meeting practical needs for handling documents, concentration and privacy. The ideal balance between these depends on the culture of the organisation and the kind of work people do.

Designing the knowledge-friendly office

'Facts' are hard to come by in this field. The research evidence shows that offices are complex, dynamic systems in which practical, personal and social needs, physical substance, symbolism and culture (and no doubt other factors) react on each other in an endless dance. The complexity and subtlety of the workplace–people dynamic means that a design that works well for one organisation can be a flop for another, and there are still no convincing, research-based rules for designing workplaces to suit different needs. However, research results and anecdotal evidence provide a number of pointers, and there is broad consensus that:

- People want – and need – to be private one minute, sociable the next. Layouts that deny this choice are unpopular and reduce productivity: too much privacy reduces casual interaction, while too little causes distraction.
- People have a deep need to personalise their workspace, to mark territory and create a home.
- Shared social spaces are most used if they are near where people routinely walk; isolated spaces tend not to be used.
- Workplace design facilitates knowledge-sharing only when the culture legitimises informal interaction; if casual socialising is frowned on, good layout cannot compensate.
- 'Magnet places' where people naturally meet (such as coffee machines) are natural centres for casual interaction.

Decade-long research by Thomas Allen and others at Massachusetts Institute of Technology into the work of product development engineers found that:

- Eighty per cent of their ideas arose from face-to-face contact; it is difficult to discuss anything complex or abstract by phone or email.
- They were four times more likely to communicate with someone 6 feet away than with someone 60 feet away – and people working more than 75 feet apart hardly ever spoke.
- Frequency of communication also depended on the extent to which people shared a common base of knowledge, the rate at which their knowledge base was developing, the size of their organisational unit, and the degree of interdependence in their work.

Allen suggested that distance might be less of a barrier in disciplines where people read more (he cited chemists), and in cultures where people are less averse than Americans to walking.

Research at University College London's Bartlett School of Architecture suggests that 80% of all work-related conversations are sparked by one person passing another's desk.

Researchers at telecoms company BT found that:

- Two people working on different floors had only a 1% chance of meeting on a given day.
- Fifty per cent of office workers regularly emailed colleagues who were only 10 feet away.

Surveys of office workers by US office furniture manufacturer Steelcase found that:

- People's paper management preferences varied – roughly equal proportions in their survey were 'neat freaks', 'pilers' or 'filers', and smaller numbers 'packrats'[1] or 'slobs'.
- Eighty-five per cent of Americans personalised their offices, and of those who did 68% said it improved their attitude at work; the most popular personalisations were photographs (69%), radios or other music players (50%), paintings or posters (47%), and flowers or plants (42%).
- Many office workers found their lighting inadequate: 56% said it gave them tired eyes, 30% headaches and 21% dry eyes. Only a minority found the overall level too dim or bright; what people most wanted was freedom from glare, and the opportunity to adjust their own lighting levels.

---

[1] Hoarders of useless objects.

Recent work by researchers at the INSEAD business school has clarified the factors that make 'magnet places' work – or fail to work – as centres of knowledge-sharing. Noting that open-plan offices and furnished circulation spaces introduced to encourage spontaneous interaction have often failed to do so, they used video to make detailed observations of behaviour in a classic location for casual conversation – the photocopier room – in the offices of a public utility, a publisher and a business school. They concluded that successful spaces share three key characteristics: opportunity to socialise and social pressure to talk; enough privacy to avoid being overheard; and what the researchers called 'social designation' – the perception that they are an appropriate and safe place to chat.

Evidence from these and other experimental studies, and from the more philosophical analysis offered by writers such as Francis Duffy of design consultancy (and workplace design specialist) DEGW, suggests that desirable features for design offices include:

- Grouping people whose work is most likely to benefit from sharing knowledge near to each other
- Arranging individual workstations on through-routes, to make casual contact possible as people walk from one place to another
- Designing workstations so that people:
  - can easily talk to both adjacent colleagues and passers-by.
  - will not be distracted by other people's phones or conversations (but limited overhearing can spark fruitful contact).
  - can see who is approaching.
  - can choose to be private or sociable, and easily indicate whether they are open to interruption or not.
  - have the flexibility to work in different ways from their neighbours without conflict.
  - can adjust their lighting and posture to suit task and mood.
  - can personalise, and even reconfigure, their workplace.
- Providing areas where people can choose to go to work as a team for a few minutes, hours or days ('dens'), with suitable surfaces, network access points, flip charts and whiteboards
- Providing easily accessible – but not too public – areas where people can have casual meetings ('clubs')
- Making 'magnet places' where people naturally have reason to wait (for a printer to disgorge or for coffee to brew, for example) attractive, easy to enter and leave, and yet offering some privacy

- Good acoustics, to reduce tiring distraction and facilitate conversation
- Consulting people to discover their functional needs and personal preferences in detail before redesigning their workplace
- Leadership that makes the culture match the symbolism and opportunities of collaborative, social offices.

Fortunately, none of these features has major cost implications, or conflicts significantly with other aims of office design. The risk involved in adopting them is small. It is all the more surprising, then, that principles that have been well documented for a decade are still honoured more in the breach than in the observance – even in design offices. And they only scratch the surface; it is clear that there are many more subtle effects at work, most of which are incompletely understood. Nevertheless, they make a practical starting point.

Workplaces for teams

Shortcomings in the design of offices tend to be blurred by habituation and adaptation, and the benefits of improvement can be correspondingly difficult to isolate from the effects of other concurrent changes. However, the effect of the workplace itself comes into sharper focus in activities that require separate companies (or just geographically separate parts of a single company) to collaborate, because there are fewer confounding factors, and the range of working arrangements is wider. Recent developments in the construction industry have created an almost ideal experiment to test some of the effects of workplace organisation, allowing direct before-and-after comparison to be made between outcomes from multicompany collaborations in which traditional working arrangements make frequent, informal communication almost impossible, and others in which modern arrangements consciously facilitate it.

Buildings are unusual among complex manufactured artefacts in being the products of collaboration between independent companies that are usually only weakly coupled and often only weakly managed by the client. As readers from the construction industry will know, the architects designing a building may have little or no contact with the engineers who design its heating and other services until a late stage in the process, and the contractors who actually build it rarely have any input into the design at all. Constructive collaboration is often further compromised by contractual arrangements that lead to confrontational positions and mutual mistrust. Predictably, the result is low productivity, high costs, time and cost overruns, a high incidence of faults, and widespread customer dissatisfaction. An investigation into the state of the industry commissioned by the UK government a decade ago suggested that as much as 30% of construction activity may be rework – just correcting mistakes. In 2006 nearly 40% of public sector construction projects were still being delivered late or over budget. All these problems can be seen as failures in learning and knowledge-sharing – and therefore in knowledge management – writ large and made visible.

To improve this dismal performance, major clients in government and industry have turned increasingly to new contract and workplace arrangements. The new contract frameworks aim to give the main parties both the opportunity and the incentive to work together in pursuit of the best possible outcome, appointing them all at the same time (or at least with substantial overlaps), making it financially advantageous for them to work closely together, and sharing risks equitably. This removes important barriers to communication, but it does little in itself to facilitate it. In addition, therefore, clients are also starting to seek further improvements by insisting that the various parties' teams should be co-located on or near site in shared offices.

A recent study reviewed experience with co-located teams and looked at evidence from a range of construction projects, and particularly closely at BAA's airport developments at Stansted and Heathrow, the refurbishment of the Bank of England, and the construction of new global headquarters for GlaxoSmithKline and the Royal Bank of Scotland – all large, complex projects. It concluded that in all these the clients

> found that co-location can greatly improve mutual understanding, dramatically speed up communication and decision-making, and lead directly to better, more buildable designs, lower costs and risk, more timely delivery, and greater client satisfaction.

None of the projects overran in time or cost, and defect rates were low. As one of the project managers said, 'You don't design everything two or three times.'

The study found, however, that not all projects with co-located teams were as successful as these. Physical proximity is not enough on its own. Mutual trust, a no-blame culture and leaders who set a strong example are as vital to effective knowledge-sharing in co-located teams as they are within organisations.

One new difficulty that arose was providing ways for professionals working away for long periods to keep in touch with colleagues in the home office, and with their practice in general. Some felt disconnected, and worried about being out of sight, out of mind and passed over for advancement opportunities. Access to the practice intranet and inclusion in circulation lists provided only a partial solution; it was the informal contact that people really missed. The gap was at least partly filled in some projects by accepting the cost of allowing people to come back to base for seminars and other work and social events. In the future, advanced video systems and the increasing availability of high-speed, high-capacity data links may help, too. Certainly, as intercompany and remote working increases, the concept of the workplace will need to be broadened out from the physical confines of an office to encompass the whole virtual space in which project teams and companies operate.

Another study, carried out by researchers at the University of Michigan, found that the improved communication within project teams that comes from co-location can pay dividends within single companies as well as in multicompany collaborations. It can be equally valuable when it simply brings together people who normally work in different locations. In a pilot project at Ford Motor Company, software development time was cut by two-thirds when the client, manager and programmers worked in one room, instead of being scattered around the company and communicating only in formal meetings or by phone or email.

These experiences with co-located teams are vivid demonstrations that frequent, informal interaction is one of the main keys to effective knowledge-sharing and collaboration. Formal meetings and correspondence, however amicable and constructive, are no substitute for the efficiency, effectiveness and timeliness of a few paces across the office and a short chat.

The same underlying factors are at work on all scales, in a single office, a single company, and in multicompany collaborations. The evidence suggests that designing working arrangements and workplaces to encourage and facilitate interpersonal interaction and knowledge-sharing pays dividends in all of them. It costs little or nothing, it has no downsides, and it works. All the organisation has to do is provide a knowledge-friendly environment and avoid negating its value with bad management practice; once that has been done, human nature takes over and the benefits flow.

Workplace design in the case studies

The Aedas Studio

Aedas created its Studio as the focus for its new emphasis on design quality. It has a variety of features designed to encourage creativity and knowledge-sharing, including new workbenches round which people can walk freely and see what their colleagues are doing, a large magnetic pin-up wall for displaying work in progress, breakout spaces for informal meetings, and a wireless network to allow senior staff to move around with their laptops.

Continued

The Studio has fulfilled expectations as a centre of creative energy, helping Aedas to win several high-profile design competitions that the practice would not previously have expected to win. It is impossible to assess how much the workplace design has contributed to this, but the indications are that it has had a positive effect. The Studio environment is well liked, and it has certainly succeeded in facilitating mutual awareness and encouraging interaction.

Buro Happold

Buro Happold has refitted its London office on the same basic principles as the Aedas Studio, with parallel rows of workbenches and pin-up walls, but with interesting features of its own. Most of the workbenches are unusually high (1050 mm, compared with the usual 725 mm), and have chairs to match, so that seated and standing people can talk with their heads at a similar level. Staff have personal storage trolleys that can be moved around to facilitate working in flexible groups, and there are large layout tables where people can gather around drawings.

Edward Cullinan Architects

Edward Cullinan Architects plan to reorganise their office on similar basic lines, with the same intention of making it easier for people to see what their colleagues are doing and encouraging casual conversation. To encourage walking around, they plan to position 'magnets' such as printers at the ends of the office rather than in the 'logical' position at the centre.

A further advantage of the new layout for ECA – as it is for Buro Happold – is that it will enable more staff to be accommodated in the same space without crowding. At ECA this will avoid the need to split staff between two offices, and help retain the coherence and sense of community they prize.

Chapter Eight
Expanding Networks

It's not what you know . . .

Ad hoc, person-to-person contact – face to face, on the phone, or by email, instant messaging or text – is by far the most common medium for sharing knowledge in most organisations, and one of the best. More often than any other process, it provides the final pieces we need to complete a knowledge jigsaw, solve a problem, and move on. But the quality of the help we get depends on who we know. The larger and more diverse a network of contacts we have, the more and better knowledge we can tap into. With only a few contacts, relying on colleagues can result in the blind leading the blind. And, unfortunately, personal networks are often surprisingly small. As the research quoted in the last chapter suggests, many people's knowledge of 'who knows what' extends little beyond the colleagues they can see from their own desks. In a recent knowledge audit in a large and successful design practice, 55% of people said they only knew what 'a few' people were knowledgeable about, even in their own office. Over a third said they knew 'nobody' in any other offices, and fewer than 3% thought they had a good idea of 'who knows what' in even 'one or two' other offices. Junior staff – the very people most likely to need help from colleagues – had even fewer contacts to draw on.

In a small organisation everyone can know everyone else, and have a good idea of what they know. They can all tap into the whole of the organisation's collective knowledge. But with many colleagues, many projects, several offices and new people joining all the time it is impossible to know what projects everybody has worked on and what special expertise they have. Much of the practice's collective knowledge is effectively inaccessible. The larger and more diverse the organisation, the more people miss out on if they only draw on a local network of contacts. And the social ecology of knowledge is surprisingly fragile: Edward Cullinan Architects (a practice of only about 30 people at the time) found that simply dividing staff between two floors noticeably reduced people's ability to know who knew what, and the ease with which knowledge flowed around. A large practice can be more like a collection of isolated knowledge villages than the vibrant city it should be, wasting one of the main commercial benefits of size. It is not surprising that, in a survey of knowledge management in 40 major companies across the world, management consultants McKinsey found that the quality of systems for identifying who knows what was a strong predictor of business performance.

We all work most of the time with knowledge that is already in our heads, but it is impossible to remember everything, and we don't try. There is no point in struggling to memorise things like tables of numbers if we can get what we want in seconds from a reference book at the back of our desk, or with a few mouse clicks. Instantly available resources such as this – our personal libraries – are effectively extensions of our own memory. If memory and personal libraries both fail to serve, the next step for most of us is to tap into our network of contacts, starting with a friend at the next desk or across the room. When that works, it is a quicker way to get an answer than to search in documentary sources. It is more effective as well when we need to know *how* rather than simply know *what*. 'Know how' is often difficult to acquire by reading, for reasons discussed in Chapter 2. But the chances of it working, and the quality of the answers, vary widely. Experienced professionals not only know more than juniors, they have bigger and better networks, too, and that adds greatly to their capabilities. Extending people's networks to give everyone access to a wider and better range of expertise is one of the easiest, quickest and most effective ways to increase organisational performance.

Despite the evidence of personal experience, many early attempts at knowledge management focused on written information and disregarded person-to-person interaction.[1] There is a seductive efficiency in the idea of getting experts to write down the best answers, once, and then using IT systems to share them throughout a company. This certainly has a part to play, and we will return to it in a later chapter. But the generally disappointing results from what has been called 'first generation' knowledge management strategies based *solely* on written information led to a recognition that much of the most valuable knowledge is either inherently difficult to capture (how to design or to lead a team, for example) or unacceptably expensive to document and keep up to date. Even when knowledge *can* be recorded in words and pictures, it is often more easily and effectively transferred in conversation. Asking someone who knows saves the effort of finding

---

[1] And few gave any consideration to whether information had any value as knowledge, either.

the right sources, searching through them, weeding out irrelevancies and translating from the general case to the particular, and the to-and-fro of conversation helps guard against misunderstanding. As the old saying has it, 'It's not what you know, it's who you know', or as McKinsey say in their survey report, 'Personal contact is the key.'

Help from IT

The value of knowing who knows what is intuitively so obvious that many professional services organisations set up databases of people and skills in an attempt to overcome the limitations of personal acquaintance even before they gave serious thought to knowledge management in a wider sense. Unfortunately, simple skills databases have often proved disappointing: little-used systems with content that is gradually degrading because nobody thinks it worth keeping up to date. Anecdotal evidence suggests that common reasons for failure include reliance on administrative staff to enter data, unfriendly inter-faces, poor search facilities, and – above all – content that is little help in deciding who to approach with a question. However, experience in Arup, Whitbybird, BP and elsewhere has shown that the basic idea of a directory of expertise is sound. It needs, though, to be based on a broader view of its purpose than simply searching for skills, promoted energetically, and above all designed with a clear focus on the *user's* point of view.

Directories of expertise deliver value only when they are in wide and frequent use. That only happens when they are easy to use (both when putting information in and when getting it out), visually attrac-tive, and above all rewarding to both grass-roots users and manage-ment paymasters. And that in turn means that they need to be seen as a rich and up-to-date source of professional knowledge, and to support management activities and the social aspects of conversation as well. They need to be useful for making up project teams, for per-sonnel management and for finding tennis partners as well as for getting answers to technical questions. This is beyond the capability of a conventional skills database rigidly controlled by management.

Software that gives users personal spaces where they can enter what they want, when they want, and which make them part of an online community and help them to expand their personal networks has proved to be much more successful. It is certainly helpful for systems like this to draw basic information from HR databases, but staff need to have responsibility for putting flesh on the bones, and write access from their own workstations: they need to see the system as *theirs*, not as management's.

'Social networking tools' like this are proliferating in the wider world. There are numerous websites that offer people the opportunity to advertise themselves on the Web and build networks of online contacts, variously dedicated to business networking, finding new friends, dating, photo sharing and a growing range of special interests. And their membership is exploding. The 'Friends Reunited' website famously developed from an experiment in the back bedroom of a suburban house into a multimillion-pound business with entries from half the households in the UK within a few years, and others have grown equally fast. The evident success of multipurpose social networking tools makes them a good model for would-be learning organisations to follow.

Social networking software can take a variety of forms. In a corporate environment, the most appropriate is usually a directory with a separate personal web page for each member of staff – a phone book on steroids (hence common names such as 'Yellow Pages' and 'Who's Who'). Networking tools like these are valuable in themselves, and they can be made even more so by interlinking them to other knowledge systems so that, for example, users can jump straight from a list of people's current projects to their pages in a project directory, and vice versa. Different people think about information in different ways, so more people will find what they want if it can be reached by a variety of routes. Links add value in other ways too. Connecting personal pages to project pages and to technical information in a knowledge base enriches all three by revealing relationships that would otherwise not be apparent, as well as simply by making it easier and quicker to find information. Searching for expertise using networking tools will not only (hopefully) show who has it, but lead to the projects where they acquired it, to other people who worked on them, to technical information on the topic, to external resources, and so on.

And that is not all. Research has repeatedly demonstrated that mutual trust is a prerequisite for effective knowledge transfer. When we trust people we are more likely to ask them questions, they are more likely to answer helpfully, and we are more likely to believe them. Mutual knowledge is an important factor in trust, and the information in a personal page can make a critical difference. Signing a knowledge base contribution (or any other document) with a link to a personal page as well as a name allows readers to look at the author's credentials, and makes it easier for them to assess the trustworthiness of the information. This in turn reduces the need for gatekeepers or moderators to guarantee quality, and the psychological barriers they create to contributions from junior staff. Personal pages can facilitate person-to-person contacts too. It is easier to cold-call someone when you know something more about them than just their 'skills', and they are more likely (and able) to reply helpfully if they can immediately call up a personal page that tells them something about you. Even professional conversations are facilitated by small talk. People recognise this: when someone in BP decided to add a link to their personal page to their email signature the habit rapidly spread round the company. Interlinking knowledge resources in ways like these can multiply their value several-fold.

In addition to their primary roles as a means of connecting people and an entry point to other knowledge resources, social networking tools can be valuable for a variety of management purposes. They can help with making up project teams, personnel management, and (by analysis of usage patterns) help knowledge managers to identify recurring issues for inclusion in 'frequently asked questions' databases, and 'gurus' whose knowledge is particularly valued. They have a wider role in knowledge management than simply as tools for discovering who knows what, valuable though that is.

One of their hidden strengths is that they connect people not just to *more* people, but to people who, by virtue of geographic or professional distance, are *different*. Thirty years ago American academic Mark Granovetter published a paper on 'The strength of weak ties', which suggested that information from people outside a person's immediate circle has particular value because it is more likely to be novel. Subsequent research has confirmed this, with evidence from a variety of fields including job hunting, the diffusion of ideas, and technical advice. A wider network of 'weak ties' can be created by extending the scope of networking tools beyond individual, current staff. IBM creates 'personal' pages for communities of practice and project teams, and at BP people leaving the company can bequeath their personal page to a colleague to keep live as a fragment of the corporate memory and a continuing point of contact. Other organisations have even invited important suppliers and other external collaborators to have personal pages. Including external contacts can considerably increase the range of knowledge available to staff, and add much more to a practice's capability than the usually modest numbers would suggest.

External networks are a particularly valuable way to tap into new and emerging ideas that can help keep a practice ahead of its competitors. It is paradoxical that research and creativity are more pervasive and embedded in professional services – not least in design practice – than in most other industries, and yet professional practices are typically less engaged in explicit research and development. Much of the R&D that an external observer might expect them to do is better accomplished in the course of practice than in separate research projects, but there are many areas where the timescales, pressures and constraints of practice, and the need for special expertise, make it a poor environment for research. Historically, research like this has been carried out almost exclusively in universities, in dedicated research and technology organisations, and by manufacturers, and practitioners have only occasionally been involved in it. The results are often published but, as we have seen, the written word usually conveys only part of the story (academic papers can be particularly opaque!). Personal contact with researchers, and ideally practical involvement in the work, are a much more effective way to tap into them. Again, contact networks provide the key.

Designing networking tools

At root, even modern social networking tools are simply databases that record information in a way which makes it easy to search and to read quickly on screen, but they are much more varied in content and presentation than conventional corporate databases. The mock-up in Figure 8.1 (prepared to focus debate about a networking directory for an architectural practice) illustrates one approach to their design that has proved successful in professional practices:

This includes sections on:

- *Contact and role information*
  - ○ phone numbers and email addresses for the page owner and secretary (where appropriate)
  - ○ roles and responsibilities in the practice
  - ○ a link to an office location map
  - ○ links for sending emails
  - ○ portrait and other photo(s) chosen by page owner
  Most of this can usually be copied automatically from management databases
- *Professional interests*
  - ○ free-text areas where page owners can list the topics they are willing to help people with, other professional interests, courses and conferences attended, and external affiliations and activities, prompted by headings that are part of the page template
  - ○ links to knowledge base pages to which the page owner has contributed
  - ○ a link to a printable CV
- *Personal interests and news* – a free-text area
- *Project experience*
  - ○ a pie chart based on time sheet bookings that indicates recent sector experience
  - ○ names and thumbnail pictures of the page owner's current projects and projects completed in the past two years, each linked to the corresponding page in the project directory. These can often be inserted automatically from management databases
  - ○ names and thumbnail pictures of other projects in which the page owner has been involved and considers

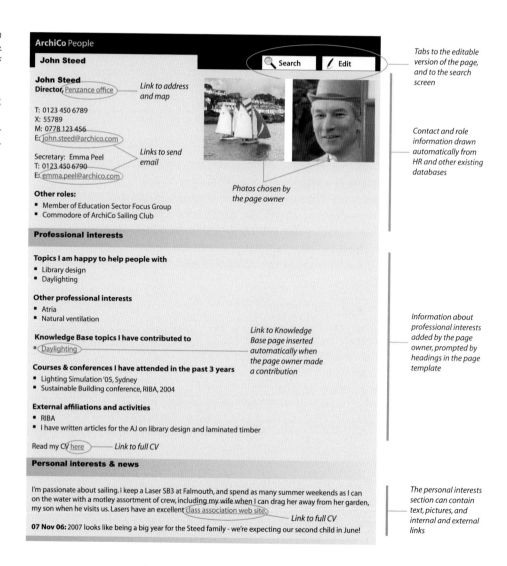

The annotations around the page read:

Tabs to the editable version of the page, and to the search screen

ArchiCo People

John Steed

Search   Edit

John Steed
Director, Penzance office — *Link to address and map*

T: 0123 450 6789
X: 55789
M: 0778 123 456
E: john.steed@archico.com

Secretary: Emma Peel — *Links to send email*
T: 0123 450 6790
E: emma.peel@archico.com

Photos chosen by the page owner

Contact and role information drawn automatically from HR and other existing databases

Other roles:
- Member of Education Sector Focus Group
- Commodore of ArchiCo Sailing Club

Professional interests

Topics I am happy to help people with
- Library design
- Daylighting

Other professional interests
- Atria
- Natural ventilation

Knowledge Base topics I have contributed to
- Daylighting — *Link to Knowledge Base page inserted automatically when the page owner made a contribution*

Courses & conferences I have attended in the past 3 years
- Lighting Simulation '05, Sydney
- Sustainable Building conference, RIBA, 2004

Information about professional interests added by the page owner, prompted by headings in the page template

External affiliations and activities
- RIBA
- I have written articles for the AJ on library design and laminated timber

Read my CV here — *Link to full CV*

Personal interests & news

I'm passionate about sailing. I keep a Laser SB3 at Falmouth, and spend as many summer weekends as I can on the water with a motley assortment of crew, including my wife when I can drag her away from her garden, my son when he visits us. Lasers have an excellent class association web site. — *Link to full CV*

07 Nov 06: 2007 looks like being a big year for the Steed family - we're expecting our second child in June!

The personal interests section can contain text, pictures, and internal and external links

interesting. For new joiners, these could include projects from previous practices
- *External contacts* – a list of organisations where the page owner has good contacts, linked where possible to entries in a contact database and/or organisations' corporate websites
- *Useful websites*, with links
- A link to a *printable CV*
- *Date of last update.*

At the top there are tabs that open the editable version of the page and the search/hit list page.

As in all websites, navigation facilities are crucial to usability. Users quickly lose patience when they are unintuitive or functionally

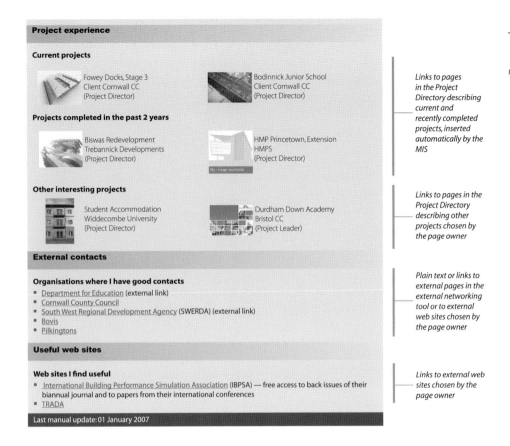

Figure 8.1 A mock-up of a page in a social networking tool (top part on the facing page and continuation above).

inadequate. Networking tools certainly need to have their own search facilities for finding people by criteria such as name, location, business unit and expertise, but in an integrated knowledge management system users are just as likely to reach personal pages through links from other systems such as a project database or a knowledge base – and vice versa – so these are equally important. How search facilities are best implemented depends on staff numbers and the nature of the content being searched. At the time of writing, architects Aedas (with over 600 staff in the UK) are experimenting with a variety of searching and filtering facilities, leaving it to staff to decide which will be retained in the final version. In a smaller practice such as Feilden Clegg Bradley, with around 100 staff, direct selection can work well for some kinds of content. FCB's *Yellow Pages* has an array of click-able thumbnail portraits on the front page and a drop-down list of names. Portraits may sound like a gimmick, but users value them. A knowledge audit at Whitbybird revealed that almost as many people 'most used' their *Who's Who*? database to find staff photos (28%) as phone numbers (31%) or information about people's role and responsibilities (29%); relatively few (7%) used it most to search for skills.

The IT infrastructure for personal pages – page templates, and the database, links to other databases and search engine behind them – needs to be set up as a practice-wide facility, but it is impracticable to have the content entered and maintained centrally. Apart from basic details, which can be copied over automatically from personnel files, content ownership needs to be devolved to the individuals concerned, encouraged by a strong example from management.

One new issue that does need to be addressed centrally is data protection and privacy legislation (and, in the public sector, freedom of information). This may become more significant in future, particularly where the scope of a directory has been extended to include external contacts; user-level access permissions can help here.

BP Amoco used an exemplary approach when it set up its system (called *Connect*) in 1997. This was intended in the first place to facilitate and encourage interaction between former BP and Amoco staff following their merger, and to replace a mixed bag of local directories. BP started by studying how other firms such as Microsoft, Glaxo Wellcome, Schlumberger and Procter & Gamble had addressed the problem of connecting people, and planned its approach carefully. It:

- Kept the design and operation of the system out of the reach of IT and HR professionals who might use it to pursue agendas that would compromise its central objective of connecting people. The only people authorised to enter and change data are page owners themselves.
- Made creation of a personal page voluntary.
- Designed data entry pages to require no knowledge of web technicalities, and to elicit key information painlessly with minimum constraint (using encouraging prompts such as 'What are you currently working on?', 'What areas have you worked on in the past?', 'What subjects might you like to be contacted about?' and 'What do you enjoy doing?').
- Encouraged people to include information about personal interests, and use photographs more interesting than passport mugshots – holiday snaps, for example.
- Persuaded the chief executive to create a personal page, complete with sections on his hobbies and interests.
- Rolled the system out first as a pilot in a small part of the organisation whose management was keen, to test the design and technology and to show others what *Connect* would be like and could do. Later, BP used focus groups to help refine the interface.

- Promoted the system energetically, using a group of volunteer 'champions', marketing initiatives ranging from technical talks to jokey competitions, success stories sent in by appreciative users, and personal touches like thank-you emails and *Connect* pens sent to the authors of the first thousand pages.

After one year about 10 000 of BP Amoco's 100 000 staff had created personal pages on *Connect*, and numbers rose to over 30 000 after four years. The company has come to regard the system as a major business asset, and it has been the inspiration for many subsequent systems in other firms.

The keys to the success of *Connect* are its connectivity and psychological sophistication – a major step forward from the impersonal, stand-alone skills databases and CV repositories that many firms still rely on. The primacy given to the users' point of view, the care over details of design and implementation, the emphasis on personal ownership, voluntariness and trust, and the encouragement to include personal information – both to facilitate professional communications and to offer other social rewards ('Are there any other dinghy sailors around here?') – have been crucial in persuading people to populate the database in the first place, and in making it a tool of continuing value. Without this personal dimension, and its links to other intranet resources and email, *Connect* might well never have gained the critical mass needed to make it a useful knowledge resource.

Philips' social networking tool (with 13 000 personal pages in 1992) is reinforced by a number of experienced and widely connected 'gatekeepers' who have volunteered to ensure responses to questions in specific fields. Enquirers can click an icon and post a query, and the system automatically forwards this to appropriate gatekeepers, who respond or arrange for someone else to do so. In the same spirit, staff at Texas Instruments can put 'information wanted' adverts on their personal pages. Bovis Lend Lease goes further in the direction Philips has taken, putting more stress on the role of human 'connectors' than on its electronic system. Their '*i*konnect' service employs a small team of full-time staff spread around the world to put enquirers in touch with people (inside or outside Bovis) who are likely to be able to answer their questions; the *i*konnect staff record questions and answers, and publish those that recur or seem likely to interest others.

Most of these systems were essentially stand-alone when they were first set up, sharing data with HR databases but with few connections to other knowledge systems. Aedas's new networking tool (temporarily called *Aedas People* – the final name will be left to staff to choose) is an evolutionary step further beyond them in its rich interconnection with the new project directory and knowledge base being developed alongside it.

Similar features tend to recur in successful networking tools, but design nevertheless needs to be tuned to suit the individual organisation. It is a moot point, for example, whether the creation of a personal page should be entirely voluntary, as at BP, or subject to at least some pressure, as at Feilden Clegg, where there are plans to use personal pages as source documents in annual appraisals, and at Broadway Malyan, where the skills field contains a default entry 'I have no skills to offer' until it is overwritten.

It is uncertain, too, how best to distinguish between different levels of expertise – important information when the system is used to search for sources of expert advice. This is not difficult in highly structured and hierarchical professions such as medicine, but much more so in others, such as architecture, where there are no universally recognised grades beyond qualification, and most people are generalists. It would be impracticably labour intensive in the latter case to make independent assessments for each separate area of expertise (though this might be feasible for top experts), so it is often left to page owners to provide the information. But if people are asked to use a simple scale such as novice–competent–expert, interpretations will differ, and anyway some people are naturally inclined to play down their abilities while others exaggerate. This undermines the whole point. Feilden Clegg Bradley have sought to get round the problem by emphasising project experience and willingness to help rather than expertise as such. This looks promising, but it is still unproven at the time of writing. Willingness to help is certainly a relevant factor (which networking tools often overlook), and approaches based on objective criteria may well prove to be the most workable. That nevertheless still leaves the problem of devising suitable criteria, which need to reflect the particular patterns of expertise and practice in the organisation.

And networking tools do not always succeed. They may fail to develop critical mass, or they may atrophy after a year or two because people stop updating their pages. The commonest cause of this appears to be lack of motivation. It can be difficult to persuade busy people to invest time now for unknown rewards later, especially when the rewards are uneven. Some people find tools like these very valuable, while others find little use for them.

Compulsion, or having administrative staff create people's pages for them, are not substitutes for self-motivation, but there are times when they are appropriate. It is reasonable, for example, to require new joiners to complete their pages as part of the induction process, because they expect to have to fill in forms, and a networking tool rewards them with instant introductions to colleagues and generally helps them get their social bearings. Later, personal pages can be used as the interface for completion of CPD records, skills audits and other mandatory processes. Basic administrative information can usefully be entered by administrative staff (perhaps via HR and other management databases) to remove the most tedious part of the page-creation process and leave less for people to add themselves. And the addition of links to other knowledge tools, such as project directories, can often be automated.

Partial alternatives to networking tools are emerging in the form of software that scans electronic traffic such as email and knowledge base searches and uses tools such as neural nets to locate interest and expertise. These are already being taken up by large corporations, but in all but the largest professional practices their high cost and the need for a large volume of communications traffic to reveal useful patterns looks likely to limit their value. The lack of a personal element is a major disadvantage, too. For the time being, networking tools are probably the most reliably valuable of the IT tools in knowledge management.

Networking tools in the case studies

The case studies include several practices that are developing their existing skills databases – which have often had disappointing usage – into fully fledged personal pages systems in order to serve people's real needs better.

Feilden Clegg Bradley

Feilden Clegg Bradley's (literally) *Yellow Pages* take advantage of its relatively small staff numbers (around 100) to take the 'human face' design even further than Arup: the home page is made up of staff portraits, each a clickable link to that person's page. It also has a conventional drop-down list of names and a search box. Pages link to relevant entries in the personnel, project and picture databases, and contain a variety of information about skills and experience. In addition to its obvious uses, FCB has given its *Yellow Pages* a key role in personnel management, using it to target CPD and select 'topic champions' – people nominated as prime contacts for technical advice. It also plans to make it a key reference in annual reviews.

Broadway Malyan

Broadway Malyan's *Who's Who* is more sophisticated, and tightly integrated with its other management tools and knowledge resources.

Each page – 'My Page' – includes:

- A summary of key and specialist skills
- Contact details
- Links to current and recent projects
- 'My contribution'

Continued

- CPD records
- 'Skills I can offer'
- 'My knowledge'
- 'More about me'
- Tools to generate a CV and appraisal report, request a business card, and report a database error
- A link to 'My time sheet'.

Buro Happold

Buro Happold has made networking the basis of its entire approach to knowledge management. It decided that 'talking to each other' was so much a part of its culture that an electronic knowledge base would be inappropriate, and it sees its networking tool as a knowledge base in itself. Launched in 2006–7, this aims to 'connect groups with groups, people with people, and skills with projects'. The main navigation interface is an expandable tree view organised at the top level in sectors, subsectors, components, Buro Happold offices and groups, people and other headings, each leading to lists that cross-reference them through links to other dimensions of the system. Each sector page, for example, links to the projects, locations, managers and experts relevant to the sector, and to documentary resources such as photos and marketing material.

Chapter Nine
Learning from Peers

Good office design and social networking tools can do a great deal to foster the informal, person-to-person knowledge-sharing that is the bedrock of organisational learning. But the contacts they promote focus largely on today's problems, and on filling small gaps in knowledge jigsaws that are already almost complete. And they depend entirely on individual initiative; management can only enable and encourage. A good knowledge strategy also looks beyond today's problems and anticipates tomorrow's. Forthcoming changes in legislation, for example, may call for new kinds of professional capability, and the approaching retirement of the baby-boomer generation may threaten a haemorrhage of invaluable expertise. Ad hoc, just-in-time knowledge sharing, however good, can do little to help meet challenges like these; an effective response requires a managed flow of just-in-case knowledge. The special qualities of face-to-face knowledge transfer (which we met in Chapter 2) are just as valuable for this, but they need to be harnessed in different ways. One of the most effective is mentoring, which twins people with a more experienced colleague as a teacher and guide.

See one, do one, teach one

Mentoring is a traditional technique for passing on both elementary knowledge about 'the way we do things round here' and high-level academic, professional and management skills. Craft apprenticeships, university tutorials and the on-the-job parts of traditional legal, medical and architectural training are all variations on mentoring, and it is used in industry to help newcomers into organisations, to develop high-flyers earmarked for top jobs, to enable experts to share their knowledge, and to help valued leavers to pass on their wisdom before they go. Superficially these could hardly be more different situations and kinds of knowledge, but they share three crucial common factors:

much of the knowledge being passed on is deeply tacit; it can only be absorbed incrementally and slowly; and it is most easily acquired through one-to-one interaction which can be tailored to suit the mentee's individual knowledge needs and their pace of learning. It is equally impossible to understand the skills involved in fine cabinet-making, analysing novels, diagnosing illness, designing buildings, understanding the culture of an organisation, or inspiring a team just by reading or listening to lectures.

It is increasingly recognised that much of the most valuable knowledge is tacit, and can be passed on effectively only by personal contact. This can be facilitated and encouraged by office design, foresight and hindsight reviews, networking tools and communities of practice, but the contacts these generate are typically ad hoc, brief and fragmented, and so is the knowledge they share. They are driven largely by opportunity and short-term need, and (with the partial exception of communities of practice) they are not capable of developing expertise in a systematic way. They deal in pieces of the knowledge jigsaw, not in complete pictures.

Conventional training courses based on lectures and reading go to the opposite extreme. They offer more or less complete pictures, but because they are based on pre-prepared material and timetables designed for an average learner they lack the personal element, and they work well only for the people they happen to suit.

Mentoring can combine the best of both. At its best, it offers mentees both a good source of just-in-time information *and* a systematic framework for developing desired capabilities and professional attitudes. In addition, it has been shown to help retain promising employees, it can help senior staff keep in touch with the grass roots, and it is one of the few practical and effective ways for departing experts to pass on their knowledge before it walks out of the door. It is the only proven technique that offers a framework for the patient, sustained, systematic and interactive communication necessary to pass on subtleties such as culture and the complexities of high-level expertise.

Given that many professionals have first-hand experience of the effectiveness of mentoring from their original training, it is surprising that it is not used more in professional services organisations. The continuing professional development that the institutions demand is more often delivered through reading and lectures. Perhaps this is because it is easier to write a cheque or arrange a few lunchtime talks than to find time for mentoring: if so, it is short-sighted.

A recent study carried out for Sun Microsystems looked at the records of over 1000 of their employees over five years to quantify the value of mentoring. This found that people who had taken part in the company's mentoring programme were at least five times more likely to have been promoted or moved to a higher salary grade than those who had not participated. The results were roughly the same for both mentees *and* mentors, suggesting that the experience helped develop the expertise of both. Retention rates were much higher for participants, too: both mentees and mentors were roughly 50% more likely to have stayed with the company than their non-participating peers. The study concluded that mentoring produced 'employees that are more highly valued by the business', and the feeling was evidently mutual. Good mentoring is financially rewarding for both organisations and individuals.

The Learning Innovations Laboratories (LILA) at Harvard investigated a wider range of benefits and identified more than a dozen, including for mentees:

- *Rapid assimilation* into the organisational culture.
- *Competence, job satisfaction and advancement.* Because mentoring provides coaching tailored to individual needs, it accelerates the development of professional capabilities, helps them to perform better, and increases their job satisfaction. LILA quote 1998 research which found that professionals who had been mentored earned $5000–22 000 more than others who had not.
- *Creativity and innovation.* Exchanging ideas with a mentor helps mentees see new perspectives and new ways to solve problems.

For mentors:

- *Enhanced self-esteem.* Being asked to be a mentor is a public acknowledgement of expertise, and mentors 'gain immense personal satisfaction from knowing that they have contributed to the development of another individual'.
- *Revitalised interest in work.* Mentees' questions stimulate the mentor's thinking.
- *Expanded awareness of the business environment.*

And for the organisation:

- *Recruitment.* LILA quote a 1999 survey of executives which found that 70% of job applicants are attracted to organisations that offer mentoring.
- *Retention.* In the year after SmithKlineBeecham introduced mentoring into a division with retention problems, losses of people who were not mentored remained high at 27%, but dropped to 2% of those mentored.
- *Cost-effectiveness.*
- *Succession planning.* Mentoring can both develop people in line for promotion or recently promoted, and help identify others with high potential.

- *Organisational change.* Mentoring can support change by reinforcing new practices and ways of thinking.
- *Increased productivity.*

Research at the Wharton School of Business and elsewhere has reached similar conclusions. The benefits of mentoring for both companies and individuals are numerous, considerable, and in some cases unique. It is a technique that professional services organisations cannot afford to ignore.

Mentoring in different contexts

The basic principles of mentoring are simple, but as with other knowledge tools and techniques the devil is in the detail. Without thoughtful management it is likely to prove disappointing.

It is self-evident that mentors need to possess the knowledge and expertise the organisation wants to pass on, and (though this is often forgotten) that they should consider personally and deeply what capabilities their mentees need to acquire. The success of all mentoring, though, depends just as much on the quality of the human relationship. Not everyone makes a good mentor. Mentors need to welcome all questions (however trivial they seem), be accessible even when busy, have the patience to repeat explanations, be happy to share their knowledge and networks, and have the communicative skill to do so and the skills to design an appropriate programme of interactions.

Since personal rapport depends on mutual trust and the fit between personalities and between cognitive and communicative styles, mismatches can happen. Both sides need to feel free to terminate the relationship without embarrassment if it fails to work. Some firms combine mentoring with formal management responsibilities, involving mentors in tasks such as staff appraisal, but there is a real danger with this arrangement that the formal responsibilities may undermine the trust needed for effective mentoring. There is some research evidence that the process works best when mentees choose their own mentors, but this is impracticable with new joiners (who normally know no one) or when mentoring is used to help experienced staff pass on their knowledge before leaving a practice. It can be difficult in other cases, too, when there are only one or two potential mentors with appropriate knowledge. In any event, mentors need to be chosen thoughtfully and individually, not only matching each mentee's knowledge needs with a mentor's ability to meet them, but as far as possible having regard to personality and psychological factors too.

Mentoring also takes time and patience. A mentoring programme needs preparation, to identify which of the organisation's needs can best be met by mentoring, and how. Potential mentors need to be identified and briefed, and when the programme has been running for some time it can be helpful for new mentors to be given opportunities to learn from more experienced peers – and from past mentees. It helps to profile mentors and mentees so that they can be matched successfully. Mentors' other workload may need to be reduced to make space for their new responsibilities, and since they may fear that this will hold back their own professional advancement, ways need to be found to recognise their effort. Mentoring for newcomers needs to be planned with particular care, both because some of it will be done by staff who are relatively inexperienced themselves, and because the learners are unlikely to be assertive in shaping the experience. And mentoring programmes need continuing commitment from senior management to ensure that they are not abandoned before they have had a chance to prove their worth, and oversight to ensure that they continue to be effective and worthwhile. It has been estimated that it takes two to three years to make mentoring a part of the culture, and for visible benefits to flow.

In the following sections we will look at the various main uses of mentoring, and the different mentors and approaches that they call for.

Mentoring new joiners

The aim of mentoring new joiners is straightforward: they become productive quicker if they are helped to learn the basics of the practice culture, office routine, administrative procedures and the other aspects of 'the way we do things round here'. New graduates also need to start developing their theoretical knowledge into real professional competence, and mentoring is one of the best ways to help them start. There can be a surprising number of practical details that are beneath the notice of university courses, but pose real stumbling blocks for people reluctant to reveal their ignorance. Long-term staff rarely need to think about basics like these, but they can seem overwhelming and confusing to the inexperienced. Driving a car becomes so automatic after a while that (at least on a clear road!) we can easily think about the next meeting and carry on a conversation at the same time, but most of us can remember how difficult driving was at first. There is just so much to remember, observe, react to and synchronise that it takes hours of patient instruction and practice to become even minimally competent, even though every part of the process is simple in itself. Joining a firm is much the same. Just as with driving, reading manuals and sitting through an hour or two of concentrated talk-and-PowerPoint is no substitute for developing competence piece by small piece through one-to-one interaction with a friendly mentor.

A mentor only a year or so senior is the best equipped to provide most of the help and guidance new entrants need, particularly when they are junior. It is hard for experienced staff to remember what they found difficult when they joined, or explain things at an appropriate level – and indeed they may never have needed to do some of the things today's juniors are expected to. People for whom the memory of joining is still reasonably fresh are more likely to have relevant knowledge and be able to empathise better, and the broad similarity in their age and stage of career provides useful common ground and makes for easier interaction. Relatively junior mentors are also, of course, less expensive and usually more easily released from other work.

A practice's core values, and an awareness of business issues, are better inculcated by a senior partner or director. They are too easily diluted and distorted in transmission for second-hand accounts to be reliable, and early misconceptions can be hard to correct. It is good, too, for new joiners to feel that they know someone at the top, and to be introduced to other seniors. Learning company-specific technical skills, such as the use of unfamiliar software, may call for a different mentor again.

Senior joiners have more relevant experience to build on (in knowledge terms, more absorptive capacity), so they can pick up the basics more easily than juniors. However, their work often requires them to interact and collaborate with a wider range of colleagues – especially if they have management responsibilities – and they become effective more quickly if they are helped to build up an appropriate network of contacts. An approach that has been found successful for this is a programme of meetings with selected people, spaced over a few weeks, accompanied (if only part time) by a senior mentor who can explain where they fit in, make the introductions, and remain accessible afterwards to help develop the network.

It is rarely possible to find mentors who can combine all these roles. The range of new joiners' distinct and different needs means that it is often best to give them two, three or more mentors, in parallel or in sequence. This sounds luxurious, but in fact it can be the most efficient as well as the most effective approach to mentoring: for each of the kinds of knowledge involved, it uses the least expensive staff who are able to do it well.

Mentoring new joiners poses a special challenge for managers when booming business is drawing in larger than usual numbers, and increasing the time pressure on other staff. There is a real temptation to cut back, but this is always a mistake. The whole point of recruitment is to add productive capacity, and the gain is more illusory than real when the result is a mass of uncertain newcomers inefficiently producing work that needs more than usual checking and correction.

Well-managed mentoring can be even more rewarding than usual at times like these.

Preparing staff for new roles

In professional practices, taking on a new role – particularly at senior levels – more often requires people to extend their management capabilities, or to develop new ones, than to enhance their professional skills. Reading and training courses have a part to play, but even today management competence still relies more on tacit than on explicit knowledge, and grows more from experience than from formal teaching. Mentoring can be one of the best ways to accelerate the process.

Shadowing the current incumbent before taking over is a well-established way of preparing for a new role in many organisations. Mentoring provides a framework for continuing the flow of tacit knowledge after the transfer, helping people to develop capability more quickly than they could if they had to work everything out for themselves. Appropriate mentors in this context include colleagues who have recently moved into similar jobs (for the same reason that recent recruits are well placed to help new joiners), other peers who are experienced in the new role, and more senior staff who have moved on. Either of the latter two can help build up networks as well as provide key information, and guide experience to develop capability. As with new joiners, it can be worth giving people more than one mentor in order to cover different kinds of knowledge. Having multiple mentors also shows new incumbents that there is more than one way of doing things, encouraging them to bring fresh ideas to bear – often one of the reasons for bringing them into the role.

Mentoring in this context – and in professional skill development, discussed below – benefits mentors almost as much as mentees. Explaining any complex knowledge is one of the best ways to clarify and develop one's own understanding, especially if the listener knows enough to ask acute questions. Mentoring a peer also provides a rare and valuable opportunity to think about issues relevant to one's own work, unburdened by personal responsibility.

Passing on high-level skills

When staff already have some – or considerable – skills and experience, mentoring can be a good way to give them more. This usually calls for a senior mentor who is more skilled and experienced all round than the mentee, but it can also be one who is simply more skilled in a particular field – even if otherwise junior. Mentoring like this – albeit not necessarily under that name – is a long-established practice in architecture, where it is common for young architects to be allocated a senior 'uncle' or 'aunt' until they have passed their Part 3 exams (as at Edward Cullinan Architects, for example). It is often policy for a senior partner to be personally involved in every job (as at Penoyre & Prasad), and there may be a designated design director with an explicit brief to develop professional skills across the practice (as at Aedas).

Mentoring to develop professional skills has much in common with the 'communities of practice' – semi-formal, self-governing networks of people who share a common interest in a specific aspect of practice – that we look at in detail in Chapter 11. They share the same principal aim (though CoPs can also have others), and both work largely through direct interpersonal contact, provide informal coaching as well as answers to questions, and have a social dimension. Each has advantages and disadvantages from a learner's point of view. Collectively, the members of a CoP often possess greater professional expertise than most mentors, and membership may offer a wider range of contacts. CoPs can also give experienced newcomers better opportunities than a mentoring relationship to feed their own special expertise into the practice and raise their personal profile. On the other hand, a CoP's responsibility is to the membership and to the practice as a whole, not to individual members, and this leaves the onus squarely on each individual to get the most out of the opportunities they provide. Not everyone has the self-confidence and assertiveness needed to build relationships that can substitute for a designated mentor. Also, by definition a CoP has a relatively narrow and usually purely technical focus, and so has little direct value as a source of help in other fields. Despite their similarities, mentoring and CoPs are complementary rather than alternatives.

Retaining expertise

Most professional practices are acutely aware of the loss of expertise and corporate capability that can occur when staff with unusual skills or long experience depart. The files and the contributions to intranet pages stay, but the most valuable knowledge walks out of the door. The only way to reduce this loss is a planned programme of knowledge transfer beforehand, and mentoring can be one of the most effective mechanisms for this. Debriefing interviews have their place, but it is difficult for either interviewer or interviewee to anticipate the full range of knowledge that could usefully be transferred, difficult to compress the process into the limited time typically allocated, and impossible to transfer the more deeply tacit knowledge through interview records. Even knowledge that has apparently been captured in debriefing interviews may well be lost for practical purposes unless it is deconstructed and woven skilfully into a well-used knowledge base, because it may never be read.

Acting as a mentor, perhaps to several staff, is a more powerful way for people to pass on their expertise before changing job, leaving a practice or retiring. If enough warning of a departure is available – and even when people resign to go to another company there is likely to be at least a month, or longer before internal moves or retirement – a mentoring programme can be set up with appropriate mentees. However, impending leavers are often exceptionally busy, and if they have resigned they may not be motivated to be good mentors; besides, it can take a long time to transfer even a fraction of a life-time's accumulated expertise. It is better practice, therefore, to involve all staff who have particularly valuable knowledge in an ongoing mentoring programme, and avoid the need for crash programmes just before they depart.

In all these contexts – new joiners, new roles, passing on high-level skills, and retaining expertise – mentoring is one of the most powerful and low-risk ways to make valuable knowledge flow more freely. It deserves to be used much more widely.

Mentoring in the case studies

Aedas design director

It has been common for years for professional practices to use mentoring as a way of introducing new staff into the organisation. Aedas is one of the few practices that applies mentoring principles to higher-level tacit knowledge.

After several years in which Aedas directors had to focus their main attention on the practicalities of merging four practices into one integrated business they became free to think more strategically again, and decided to make improving design quality their first priority. Recognising that high-level design skills are largely tacit, they created a new post of design director to lead the process, and set up the Aedas Studio to provide a supportive environment for it.

The design director's role is to lead by example and act as a mentor for architects working in the Studio, working with them every day and using the jobs going through the office as vehicles for developing their design skills. A key part of the vision behind the initiative is that the architects in the Studio should have free access to and sustained contact with him. Aedas believe this will generate substantially greater benefits than spreading his time across their nine offices, where he could be only a relatively remote and occasional influence – a conviction that is a key part of the rationale for mentoring. They plan that architects from the other Aedas offices will come to work in the Studio for specific jobs so that they can also benefit from the process.

Chapter Ten
Learning from Practice

Knowledge-friendly office design, extended networks and tools that help people discover 'who knows what' can do much to restore the natural flow of knowledge around organisations that dwindles so rapidly as they grow and fragment. They all help to break down silos, bring the organisation's collective knowledge into more and wider use, and raise standards everywhere towards the best. But in a competitive world today's best is unlikely to be good enough next year, and improvement can come only from *new* knowledge, embodied in new products and ways of working. Sharing existing knowledge better is not enough; a knowledge strategy needs to encourage and facilitate creativity and innovation, too.

There is no hard line between sharing existing knowledge and creating new; what is new to one person is often old news to someone else. Most of the tools and techniques used in knowledge management contribute to both. Sharing existing knowledge both multiplies its value and leads to the new conjunctions of ideas that fuel creativity. Relatively little major innovation in professional practice stems from genuinely new understanding created in-house. Most of it is the result of adopting ideas that originated elsewhere, and simply adapting them to local circumstances and needs (knowledge management itself is a case in point). Ideas from outside often come in at a single point – through one person reading, attending a conference, or talking to someone – and then knowledge-sharing is fundamental to deriving business benefit from them. However, even simple knowledge that originates in-house can be disproportionately valuable, because it can be unique. Knowledge from outside is usually equally available to other organisations, so it is a source of competitive advantage only to the extent that one adopts it earlier and uses it more effectively than another; eventually, everyone else catches up. Competitors have no access to knowledge created internally until they discover it independently themselves, or it gets into the public domain.

Practice: the invisible lab and unsung teacher

Few professional services organisations carry out original research as such, and the source of most new insights is the invisible laboratory of day-to-day practice. New knowledge bubbles up unpredictably all over, most often in small pieces – practical 'lessons learned', rather than complete new theories or ways of working. The gradual accumulation of lessons like this is the main factor that turns new graduates into senior professionals, and on a larger scale it is what makes organisations unique and provides a major source of competitive advantage. Many of the lessons, of course, are new only to the individual learner. Others are potentially of wider value, but in the normal course of events they spread only through chance conversations, and not infrequently they do not spread at all. And people rarely learn as much as they could from their experience, because they do not take the time needed to think through its implications, collect the evidence that could harden suspicions into certainty, or compare perspectives with the other people involved. Some lessons are obvious, but the real cause of events – which is the important thing to know, not the simple fact of the event happening – is often unclear, and it may not be knowable without input from others. This creates three challenges for the would-be learning organisation: to help people learn more from their day-to-day experience; to record the lessons that have widespread value in a way that makes them available to everyone; and to make sure they are used.

The benefits of learning more and making better use of the lessons are most obvious when things have gone wrong. As Ove Arup put it, 'Mistakes are valuable guides. They should not be forgotten or concealed.' All organisations are prone to reinventing wheels and repeating mistakes, and professional practices are not immune. Architects typically report in knowledge process audits that they waste 15–20% of their time reinventing wheels (many guess far more), and this is largely a failure of organisational learning. In organisations where staff time is overwhelmingly the largest cost, reducing reinvention on this scale by just a quarter translates into a 4–5% cut in the total cost base – and a much larger increase in profitability. Mistakes can be just as damaging financially as duplicating effort. One of the strongest criticisms of the British construction industry made in a government review led by Sir Richard Egan (at the time Chief Executive of airport operator BAA, one of the largest construction clients in the UK) was the high incidence of mistakes, which Egan blamed for driving up costs and leading to unsatisfactory quality and widespread client dissatisfaction; again, this is largely a failure of learning.

Problems like these are not unique to construction. It has recently been estimated, for example, that medical errors are the fifth most common cause of death in the USA, leading to nearly 100 000 avoid-able deaths annually. Across the developed world the average odds of dying as a result of medical error while in hospital are one in 300, 33 000 times higher than the risk of dying in an air crash. It is some-times argued in defence of professional services that they are more varied than most other business activities, but in fact few construction projects, medical cases or consultancy assignments are genuinely unique; most of the things professionals do are variations on things they or others have done before. Most of the reinvention and many of the mistakes could be avoided if individual practitioners, and organ-isations, learned all they could from their experience of day-to-day practice.

The benefits of learning from *successes* – the 'best practices' that organisations all seek to propagate – are too widely recognised to need reiteration here.

Windows of opportunity

The best time to reflect on experience and learn lessons from a project, whether a building or a treatment programme, is immediately after it is completed, when the pressures are over and before memo-ries fade – particularly if there have been problems.[1] The US Army was one of the first organisations to recognise this, and in the 1980s they developed a systematic process for post-project learning. They called it 'After Action Review', and this has become the generic term for post-project learning, especially in America. The oil company BP adopted essentially the same process – with considerable success – and called it a 'retrospect meeting', or 'Learn After', and many others have followed. After Action Reviews and BP's retrospects are essen-tially workshops with a particular structure, and while this is undoubt-edly the most widely useful approach, there are others. I call them all, collectively, *hindsight reviews*, and they can all be useful frameworks for learning lessons, sharing them with the whole project team (and sometimes a few others), and recording them for wider dissemination.

Hindsight review and individual learning are not the only sources of new knowledge in professional practice. The very beginning of proj-ects, the only time when there is space for new ideas relatively free

---

[1] A second good time to review experience is some time later, when the final outcome from the project is revealed – when a building has been occupied for a year or two, a consultancy assignment has been completed, or a hospital patient has survived a reas-suringly long time. Later reviews like these have proved very valuable in construction (where they are usually called 'post-occupancy evaluation') and medicine, but they are more often carried out by third parties as a research exercise than by the original project team, and as such they are outside the scope of this book.

from the constraint of prior decisions, can provide an opportunity for more fundamental creativity and learning. In many projects, of course, there is an obvious way forward, and even the initial space is too small to offer much scope for creativity. But when there is no obvious path, and particularly when objectives are challenging, lateral thinking and radical innovation are possible – indeed, they may be necessary for success. One of the best ways of enabling this has been found to be a workshop that brings together the core project team and an appropriate selection of specialist experts for an intensive effort over a day or more, followed up by rigorous analysis of the ideas that emerge. A workshop like this provides the necessary concentration of knowledge and experience, and – with good leadership – meeting face to face, away from the interruptions of normal work and with only a short time available creates an intellectual buzz that no other approach can match. Like hindsight, this process – which can conveniently be called *foresight* – has proved to be highly successful. The BP/Bovis Global Alliance, for example, has used foresight twice to find innovative ways to cut the cost of building service stations: on the first occasion they slashed it by over 25% (saving $74 million in the following year alone), and when they repeated the process a few years later they cut a further 30%.

Together, foresight workshops and hindsight review offer a systematic and proven approach to in-house learning in professional practices. They excel in harnessing the collective knowledge of groups, tapping into the tacit knowledge that exists only in people's heads, and in exploiting the magic of face-to-face communication and the intellectual buzz of engaged teams. From a knowledge management point of view, they have the practical advantages of involving little or no investment: each event is a one-off, with its own pay-off. They both deserve to be much more widely used in professional practice.

Foresight: learning from invention

Foresight is simply a systematic process for bringing existing knowledge and expertise to bear more effectively on new projects. It focuses on tacit knowledge, because that is where most of a practice's memory of previous wheels and old mistakes is stored – together with insights into how to make better wheels and avoid the mistakes – and because tacit knowledge is so often under-exploited. Equally importantly, foresight provides a forum for interaction between fresh and experienced minds, and that can be intensely creative.

The foresight process is based on loosely structured discussion between a new design team, colleagues and occasionally outsiders who have directly relevant experience. A foresight workshop typically involves 4–8 people, and it can last from an hour to a day or more, depending on the scale of the project. It is held right at the beginning of the project, before any decisions have been made that could fetter creativity.

Key steps include:

- Deciding on the *focus* for the workshop. It could concentrate (for example) on aspects of particular concern, or in which the team is inexperienced. The clearer its purpose and (within reason) the narrower its focus the more likely it is to be productive. In an architectural practice, for example, 'design', or even 'school design', are probably too broad; a topic such as 'making natural ventilation work in a deep space' is usually better.
- Deciding on *objectives*. Do you want the workshop to produce an outline scheme, identify specific products or suppliers, or simply point at relevant project files and exemplars? Is there a specific target for improvement on previous cases?
- Deciding on the *scale* of the workshop: should it be squeezed into a lunchtime, take a whole afternoon, or spread over a couple of days? This will depend on the focus and objectives, and on factors such as the value of the project, how novel it is (for the project team), and the perceived risks. It is false economy to make foresight workshops too short: it is important for people to have time to develop and articulate their thoughts and for interesting issues to be talked through, while breaks for reflection and socialising can make the discussion much more productive. To avoid the temptation to penny-pinch, and to ensure that foresight never degenerates into an unproductive ritual, the process should be reserved for the minority of projects that can benefit from a fresh and innovative approach.
- Identifying the *people* with the most relevant skills and experience. Social networking tools are invaluable for this, but asking senior colleagues who have been with the practice for a long time can help too. One of the disadvantages of too broad a focus is that it becomes difficult to narrow the choice. Trust is vital to free discussion, so it is sometimes necessary to take personal compatibility into account, too.
- *Briefing* the participants on the project and the purpose of the workshop a few days in advance, to give them time to collect their thoughts.
- *Assembling* resources such as books, project files, journals and photos that can illuminate the discussion: foresight concentrates on tacit knowledge, but documentary evidence can make important contributions too.

- During the workshop, *guiding* the discussion to make time for the issues in the team's mind to be articulated; for other participants to describe their experiences and offer their insights; for the issues that emerge as most important to be debated in enough depth to give useful results (there may not be time to debate them all); and for conclusions to be summed up and reviewed at the end.
- Making a *record*. Most people like to take their own notes, but when discussion becomes intense it can be impossible to keep up. It is worth making an audio recording and photographing flip charts so that people can check back on points they missed. Exceptionally, it may be worth having a rapporteur, too.
- *Following up* the ideas generated. The project team and experts need to review the ideas and their implications critically. They may turn out on examination, for example, to offer less performance benefit than expected, to be mutually incompatible, or to be too expensive, and when this happens either they will need to be developed further or a search will need to be started for a new solution. This may call for a second workshop.

Practicalities

The reductions in wasted time, mistakes and risk, and the incremental improvements in design quality, that foresight routinely achieves can give a good return on the time invested. It can, though, achieve much more.

Psychiatrist (and cybernetics pioneer) W. Ross Ashby suggested 40 years ago that a useful distinction could be made between what he called 'single-loop' and 'double-loop' learning. In recent years the idea has been developed by the leading organisational learning thinker Chris Argyris and others to become a key concept in knowledge management theory (Figure 10.1).

Single-loop learning seeks to make improvements within the boundaries of conventional thinking; double-loop learning challenges the conventional thinking, and breaks through the boundaries to find radically better solutions outside. It is single-loop thinking, for example, to make offices more energy efficient by using more efficient luminaires; it was double-loop thinking (some years ago!) to realise that energy demand could be reduced much more by designing for higher levels of daylight and using just enough electric light, under photocell control, to make up design illumination levels. And that in turn reduces the need for chillers . . . Foresight workshops provide an ideal environment for this kind of thinking.

The Global Alliance's second approach to foresight used a three-stage workshop process, including ideas from value engineering:

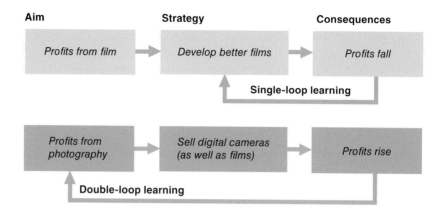

Figure 10.1 Single-loop and double-loop learning.
After decades of growth, companies such as Kodak and Fuji saw film sales and profits fall sharply when digital cameras started to become popular. A single-loop response would have been to do what they had always done in the past when their market share slipped: develop better films, and cut prices. A double-loop response challenges the premise that the company makes its profits from film and replaces it with the vision that it makes its profits from *imaging*, of which film is just one example. This opens up other possibilities such as making digital cameras and personal photo printers, and offering in-store and web-based printing services – just what Kodak and Fuji have done.

1 Reviewing functional requirements and information on cost and performance in previous projects, identifying areas where there appeared to be scope for savings, and developing criteria for evaluating solutions. The Alliance brought information and expertise from all its teams worldwide into the process: their eventual solution, for example, included the use of a Portuguese company to supply and install largely prefabricated furniture and equipment in service station shops.

2 Looking more closely at the functional requirements of focus areas and brainstorm alternatives. In 2002 the Alliance team used two workshops for this. After the second, members of the team took on responsibility for working up detailed solutions for their various areas of speciality, based on the ideas they liked best.

3 Reviewing the worked-up solutions and evaluating them against the criteria set in stage 1.

In the search for radical improvement, the Alliance stress the importance of focusing on functionality – what the system needs to do, rather than its physical nature – as a way of detaching thinking from past solutions and stimulating creativity.

At root, the success of the Alliance's foresight process is the result of making much more effective use of their corporate knowledge resources than a conventional design approach normally achieves. They emphasise the importance of supportive underlying conditions: a culture of learning and knowledge-sharing supported by visible commitment from management, appropriate incentives, and good processes. Bovis Lend Lease is rich in these, including communities of practice that pool the expertise of groups of experts in specific areas from around the globe; iknow, a database of research; written reports and knowledge networks across the organisation; and ikonnect, a service that uses facilitators based in London, Sydney and New York to find answers to questions by putting people in touch with expertise elsewhere in the company. Their success in using foresight to cut the cost of building service stations suggests that their wholehearted commitment to knowledge management is paying off.

Hindsight: learning from mistakes – and success

The value of looking back at completed work and learning lessons is self-evident, and many design practices include post-project 'close-out' reviews in their quality procedures. The common experience, though, is that in practice they either rarely happen or they degenerate into a cursory exercise in box-ticking. In either case, the reports are little read; getting on with a new project is always more appealing than doing an unrewarding chore. This is a missed opportunity: potentially valuable lessons are forgotten, not shared, or never learned. Experience in many industries has shown that a well-designed hindsight process *can* work, producing tangible increases in professional skills and process efficiency while being personally rewarding for the participants as well.

The shortcomings of conventional close-out reviews are not hard to find. People are reluctant to carry them out because they know they have never found much value in the review archive, so the effort seems futile. There are no other significant rewards, and sanctions for non-completion are often non-existent; when they exist, the obvious tactic is to do the minimum that will satisfy the bureaucracy. The reports stay unread for a variety of reasons. With the emphasis on producing a piece of paper rather than on reflecting and learning lessons (which takes time), they rarely contain information of much value. They are not written with future readers in mind, and at worst they are merely ticked boxes and platitudes. Potentially, many of the most valuable insights would come from recognising mistakes and the possibility of improvement, but the psychological temptation to sanitise the story and create a comfortable memory is often irresistible: confident assertion of success is more likely to be rewarded than admission of shortcomings. Inevitably, the results are typically short of credibility, content and interest.

Hindsight systematically addresses the shortcomings of conventional close-out reviews. It costs more, but used selectively (as it should be) it provides real value. There are two main differences in the process:

- Hindsight separates gathering the facts about what happened from extracting lessons and sharing results. This makes sure that each step receives due attention, helps develop good habits of reflection and learning, keeps the purpose of each activity clear, and is more effective all round.
- It is a group effort, involving all the principal actors in the project; ideally, these include outside collaborators such as clients and contractors, and key levels below management. Experience shows that many of the most valuable insights into the reasons for successes and difficulties come from juxtaposing multiple perspectives and understanding why they differ. This brings to the surface understanding from the deepest layers of the knowledge iceberg that we met in Chapter 2 – unconscious tacit knowledge that only the group as a whole possesses. As philosopher David Hume put it, 'Truth springs from argument among friends.'

Hindsight is most often based on semi-structured discussion in a workshop – as in the classical After Action Review – but it can also be based on individual interviews, or a mixture of the two. Group discussion can be more productive, with one speaker sparking ideas in others; it also helps to develop networks, and provides social rewards. Interviews, on the other hand, score by avoiding the difficulty of gathering busy people together in one place and time. They generally allow more input from each participant, and they can be particularly useful if there is a risk of strained relations inhibiting frank discussion. The overall cost of the two approaches is usually broadly similar.

Workshops

Hindsight workshops are quite different from conventional project management meetings or project reviews. Their purpose is reflection and learning, not making decisions, persuading people, or self-justification, and the focus is on significant events and issues; routine events are ignored.

Many organisations that use hindsight reviews find it useful to have some formal 'rules of procedure' (see box below). Until everyone becomes familiar with the process, it is helpful for the workshop leader to start by reminding participants of these rules and the aims and structure of the event. It is important for them to understand that:

- A hindsight workshop is a candid, non-judgemental discussion of what went well and what went less well in a project, intended to help everyone present – and other colleagues – do better in the future. Contributions will not be individually attributed in any report, and nothing anybody says will be held against them in the future.
- Everybody's contribution is equally welcome and potentially valuable; everybody is encouraged to contribute, but nobody is obliged to do so.
- Contributions should focus on personal knowledge. Objective facts, personal perceptions of events (even if subsequently found to be factually inaccurate) and the thinking behind decisions are all equally important. Nobody should speak on another's behalf, and speculation about other people's perceptions should be avoided.
- It is normal for people's views of events to differ: the differences often reveal where performance could be improved. There should be no attempt to find out 'who was right'; normally, all views are legitimate reflections of the circumstances of the original experience.
- Criticism must be avoided; equally, everyone should wear a 'tough skin' and avoid interpreting as criticism perspectives that happen to conflict with their own.

Workshop rules of procedure

- Nobody is required to speak, but everyone is strongly encouraged to do so.
- All participants have equal status during the workshop.
- Everybody speaks only about their personal experience in the information-gathering phase.
- Everyone recognises that subjective truth can differ from person to person.
- Nobody criticises anyone else – the focus is on past truth and future improvement.
- Management guarantees no recriminations.

Preparation for a hindsight workshop is broadly similar to preparation for a foresight event, but there are important differences:

- The *objective* of a hindsight workshop is simpler than in foresight – to learn useful lessons – but the focus is more diffuse. Experience suggests that it is best to concentrate on notable events: unexpected successes or technical difficulties, and shortcomings such as serious cost or time overruns. Defining the focus involves reviewing project records to identify critical issues such as these, and may also involve contacting key people to help identify significant events and issues.

- The *scale* of the exercise (of which the workshop may only be part) will depend on the likely value of lessons in future projects. A simple checklist that takes into account factors such as the number and significance of the notable events and the likelihood of similar circumstances arising again in the future will help project leaders decide whether to carry out a review and, if so, how much it is worth investing in it. Hindsight is worthwhile in a higher proportion of projects than foresight, but it too should be used judiciously; it is better for people to want more than for them to be disillusioned by a series of unproductive events.

- The *plan* for a hindsight workshop can be based either on project chronology or on notable events. It is helpful to summarise these on slides or flip charts as a visible reference. When time is limited, immediate focus on the notable events identified in the initial review can be the more productive approach, but when time allows, a chronological structure is preferable because it allows new notable events to emerge.

- When projects involve *other organisations*, key people from their teams should be invited to take part. Misunderstanding and poor communication between organisations is a common cause of problems, and, without representation from all parties, causes – and hence solutions – can be impossible to discover. However, the group dynamics of a workshop change when numbers rise above ten or so. At its best, discussion in smaller groups is naturally lively, interactive and usually productive, but above ten it either becomes more formal or degenerates into a verbal brawl: both reduce effectiveness. When there is a larger number of important contributors, it is better to interview some or all individually, either replacing the workshop entirely or restricting it to a manageable number.

- Hindsight reviews need a *leader* to identify the notable events in the project that are most likely to reward investigation, who should take part, and whether the process should include interviews; to collate a project timetable and other key documents for later reference; to carry out the interviews if there are any; and to record lessons learned in the corporate knowledge base in a form that non-participants will find useful. Leaders need to be familiar with the project and its business context, but they also need to be independent of the project under review: it is almost impossible to be detached about your own work. Leading a hindsight review is an excellent learning experience in itself, and it is good practice to make the role a recognised step towards promotion. It should certainly never be given to the person who can most easily be spared from other work.

- Workshops also need a *facilitator* to ensure that the plan, the timetable and the rules of procedure are followed, and that issues previously identified as important are discussed, to prompt for the reasons behind actions, and to control the overly voluble and encourage the shy. This is a full-time job during the event, and it is difficult to combine it effectively with overall leadership of the review. It is worth employing an experienced outside facilitator for early workshops, until in-house staff become familiar with the procedure and develop the necessary skills. Not everyone has the right skills or personality, so in-house candidates should be chosen with care.
- *Briefing* is more important than in foresight events when there are outside participants unfamiliar with the process. It should include an explanation of what the hindsight process aims to do and how it works, including the workshop rules of procedure, and ask invitees to refresh their memories of the project before the workshop.

The classic After Action Review is divided into three main phases, taking around 25%, 25% and 50% of the time respectively:

1 *What* happened?
2 *Why* did it happen?
3 *How* can we do better?

This remains a good framework both for hindsight workshops and for interviews. In practice, the cycle may be repeated for each of the notable events.

In phase 1, the objective is to establish what was *supposed* to happen and what actually *did* happen – what the US Army calls 'ground truth'. This provides a solid foundation for subsequent discussion, and gives participants a clear, shared view of the interaction of people and events in critical parts of the project. It is helpful to have project records available for reference.

The objective in phase 2 is to discover *why* significant events happened as they did. Research by Gabriel Szulanski with the American

Productivity and Quality Center showed that one of the main barriers to the spread of best practice (perhaps the largest barrier in a professional practice context) was what he called 'causal ambiguity': uncertainty or misunderstanding about the real reasons for success and failure. This is all too common. As we saw in Chapter 2, the human mind is almost irresistibly driven to seek causal explanations for events, but it is relatively unconcerned about truth; it focuses on immediate causes, and it is satisfied with simple plausibility. This leads people to seize on the first simple explanation that fits their mental model of the world. In an organisational context, the result is that people are apt to attribute successes entirely to their own brilliance and skill when in fact these are only part of the story, or even entirely irrelevant, and attribute failures to others. Researchers at the Massachusetts Institute of Technology (MIT), for example, found that a delay blamed by a product development team on an external supplier was actually caused by their own purchasing department's failure to pass on crucial information. Applying the 'lesson' that the supplier was unreliable and should not be used again would have achieved nothing (and unfairly penalised a good supplier); the real lesson was that the purchasing department needed more staff.

Misattribution like this is rarely a conscious attempt to claim undeserved credit or deflect blame; more often than not people actually believe their explanations, misled by their happily self-deluding minds. But reproducing misunderstood 'success factors' is a recipe for disappointment, as Szulanski found. Causal ambiguity leads to ideas being (mis)applied in situations where they cannot work well, and it can easily lead to disillusionment with knowledge-sharing and a retreat into mental silos. The aim in phase 2 of hindsight reviews is therefore to discover the real, root causes of events. This requires the patience and determination to trace events back to specific, unforced actions, and to identify any contextual factors that influenced the outcome. The main technique for doing this is repeatedly asking 'Why?' until there are no more whys to be asked, and it is one of the facilitator's main responsibilities to ensure that this is done.

Phase 3 is where lessons are learned. The discussion should build on the results from phases 1 and 2 to identify where improvements can be made, and how. The aim is to explain as clearly and explicitly as possible what should be done, and include any caveats (about applicable context, for example), warnings and uncertainties. When that is impossible, it is still helpful to give clear pointers to where solutions are likely to lie.

Interviews

Interviews come into their own when it is difficult to get everyone together for a workshop, or when there are too many important contributors for a workshop to be successful. They follow similar lines to workshops, but they can probe more deeply in some respects into 'what, why and how'. Project participants can reveal much more in a 40 minute interview than in a 3 hour workshop shared with a dozen others, they are less likely to be diverted from lines of thought, and as the undivided focus of attention they can be prompted more thoughtfully. They may also be less inhibited – but they will not have their thoughts sparked by other participants' contributions.

MIT's Center for Organizational Learning used interviews to collect evidence in a series of large-scale, multi-month reviews looking back at major events and projects. (One, for example, was the development and launch of a new car.) They found that it was helpful for interviewers to work in pairs, with an 'insider' able to recognise and ferret out critical details, and an 'outsider' free to ask naive questions and raise 'undiscussable' issues that the insider might avoid: this also eases the note-taking load and gives interviewers more time to think. This is equally worth considering in more modest reviews.

Analysing results and sharing lessons learned

A hindsight workshop can be an all-in-one activity, learning lessons and sharing them – if only among the participants – within the workshop itself. When interviews are used, the different perspectives can only be compared and lessons extracted after all are complete. In both cases, wider sharing of lessons learned requires thoughtful documentation.

Extracting lessons from interview records involves juxtaposing stories with each other and with project records to trace notable events back to their underlying causes. It may be necessary to refer back to interviewees to fill gaps as the analysis proceeds. The results often point directly to ways to avoid problems and do better in future, but with only one analyst involved, rather than six or more workshop participants, there is less scope for creative thinking. Further analysis may also be necessary after workshops when trying to record lessons learned reveals that causes remain uncertain. When that happens, the choice is to discard the lesson (reasonable if it is only minor), record it with a prominent health warning, or investigate further with additional interviews and evidence from project records in order to trace causality with confidence.

The appropriate medium for sharing lessons learned depends on their nature and importance. Typically, most are specific to a particular situation: in construction, for example, they may be associated with a particular detail of design, material or process. The ideal medium in most cases is perhaps a knowledge base, where all the lessons on a particular topic can be collected together in one place, with links from each to a corresponding project page that gives further information about its context. Juxtaposing related lessons together in this way encourages people who contribute new lessons to consider their relationship to others recorded previously, and comment when they confirm or appear to contradict them. Most importantly, it also ensures that anyone seeking information about a topic is presented with all the relevant lessons. It can be helpful to duplicate the lessons in the appropriate project file, to help people who look there rather than in the knowledge base. Some will make good anecdotes and stories for a house magazine. They will reach a wider audience there, and research shows that personal stories often transmit knowledge more effectively than dry technical statements and bullet points. A few of the most important may justify amending corporate guidance notes or procedures.

The medium that is least often appropriate is a stand-alone document, whether in the form of a page of bullet points or (even worse) of workshop minutes. Documents like this commonly disappear into a file store, never to be read again. Stand-alone records do, however, have a place in certain circumstances. MIT developed a special format, which they called a 'learning history', to help communicate the results of their large-scale reviews to the many, geographically dispersed people who had been involved in the original projects, and to others who could benefit from studying MIT's conclusions. Each chapter in a learning history reviews a particular event and is introduced by a full-width column explaining the basic facts and their business significance. Below that, a narrow left-hand column gives the learning historian's commentary, reflections and insights, designed to provoke readers into deeper thoughts, and the right-hand column contains verbatim quotations from (anonymous) interviewees, revealing their individual points of view. MIT found that the use of personal voices was a great help in making lessons learned more understandable, memorable and credible – characteristics they share with other kinds of storytelling (discussed in Chapter 12).

It can be worth writing learning histories even in situations where they are inappropriate as a dissemination medium. Buro Happold and Edward Cullinan Architects have both found that the format makes a good framework for organising perspectives on the 'ground truth', and provides a helpful mental discipline for extracting lessons. Figure 10.2 shows a page from Buro Happold's report on one of their hindsight reviews.

Despite the success of learning histories, MIT did not rely entirely on them to disseminate their findings. They also organised knowledge-sharing workshops where the results were discussed, using the learning histories as a reference, and encouraged as many people as possible to take part. Workshops gave participants an opportunity to engage actively with the lessons and the experience that gave rise to

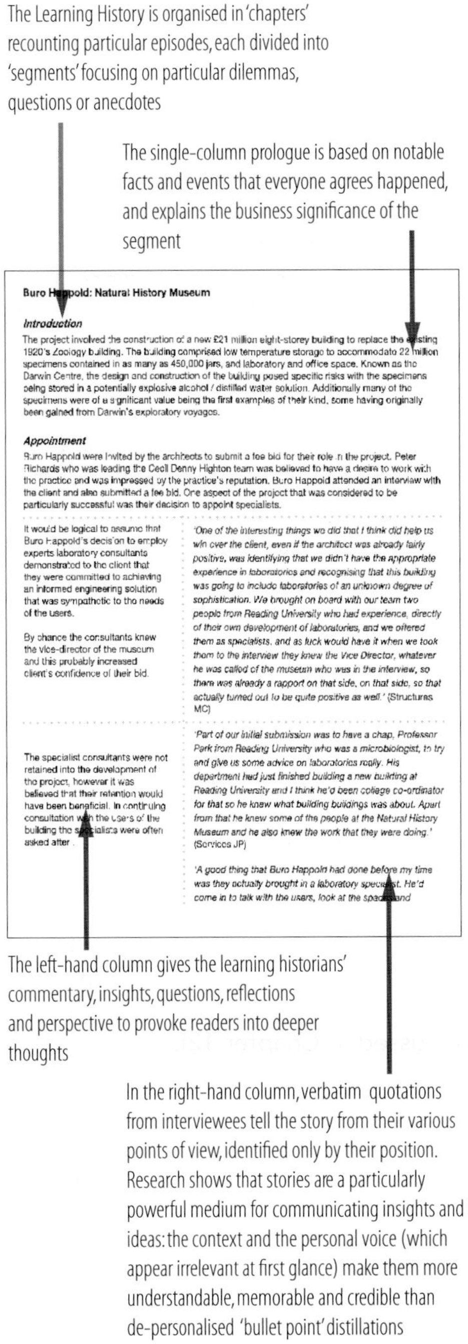

The Learning History is organised in 'chapters' recounting particular episodes, each divided into 'segments' focusing on particular dilemmas, questions or anecdotes

The single-column prologue is based on notable facts and events that everyone agrees happened, and explains the business significance of the segment

---

Buro Happold: Natural History Museum

Introduction

The project involved the construction of a new £21 million eight-storey building to replace the existing 1920's Zoology building. The building comprised low temperature storage to accommodate 22 million specimens contained in as many as 450,000 jars, and laboratory and office space. Known as the Darwin Centre, the design and construction of the building posed specific risks with the specimens being stored in a potentially explosive alcohol / distilled water solution. Additionally many of the specimens were of a significant value being the first examples of their kind, some having originally been gained from Darwin's exploratory voyages.

Appointment

Buro Happold were invited by the architects to submit a fee bid for their role in the project. Peter Richards who was leading the Cecil Denny Highton team was believed to have a desire to work with the practice and was impressed by the practice's reputation. Buro Happold attended an interview with the client and also submitted a fee bid. One aspect of the project that was considered to be particularly successful was their decision to appoint specialists.

It would be logical to assume that Buro Happold's decision to employ experts laboratory consultants demonstrated to the client that they were committed to achieving an informed engineering solution that was sympathetic to the needs of the users.

By chance the consultants knew the vice-director of the museum and this probably increased client's confidence of their bid.

'One of the interesting things we did that I think did help us win over the client, even if the architect was already fairly positive, was identifying that we didn't have the appropriate experience in laboratories and recognising that this building was going to include laboratories of an unknown degree of sophistication. We brought on board with our team two people from Reading University who had experience, directly of their own development of laboratories, and we offered them as specialists, and as luck would have it when we took them to the interview they knew the Vice Director, whatever he was called of the museum who was in the interview, so there was already a rapport on that side, on that side, so that actually turned out to be quite positive as well.' (Structures, MC)

The specialist consultants were not retained into the development of the project, however it was believed that their retention would have been beneficial. In continuing consultation with the users of the building the specialists were often asked after.

'Part of our initial submission was to have a chap, Professor Park from Reading University who was a microbiologist, to try and give us some advice on laboratories really. His department had just finished building a new building at Reading University and I think he'd been college co-ordinator for that so he knew what building buildings was about. Apart from that he knew some of the people at the Natural History Museum and he also knew the work that they were doing.' (Services JP)

'A good thing that Buro Happold had done before my time was they actually brought in a laboratory specialist. He'd come in to talk with the users, look at the spec and

---

The left-hand column gives the learning historians' commentary, insights, questions, reflections and perspective to provoke readers into deeper thoughts

In the right-hand column, verbatim quotations from interviewees tell the story from their various points of view, identified only by their position. Research shows that stories are a particularly powerful medium for communicating insights and ideas: the context and the personal voice (which appear irrelevant at first glance) make them more understandable, memorable and credible than de-personalised 'bullet point' distillations

Figure 10.2 A page from one of Buro Happold's learning histories.

them, and to make them their own in a way that simply reading a report or (even worse) listening passively to a presentation usually fails to do. Arup have found knowledge-sharing workshops valuable, too, for similar reasons.

Workshops can be an excellent way to disseminate lessons, but they are expensive unless the target audience is relatively small and geographically compact. It is rarely possible for all the relevant audience to attend; staff change; and people forget. A documentary record of some kind is indispensable to getting maximum value from hindsight, however it is carried out.

Choosing cases

Foresight and hindsight are not appropriate for every project. There is little value in foresight when a project is straightforward and the design team is experienced, or in hindsight when a routine project has gone according to plan, with no good or bad surprises. In a construction context foresight is most likely to be worthwhile when:

- A project presents unusual design, cost, time or client relationship challenges, and
- Other members of the practice have more relevant experience than the design team.

Hindsight is most worthwhile when:

- A project has over- or underrun significantly in time or cost.
- The design changed significantly more than usual between sketch and final design, or during construction.
- More projects of a similar kind are in prospect.

In either case, the effort invested in a review should reflect its likely value, and it is worth developing rules of thumb to help decide whether to carry out foresight or hindsight and, if so, on what scale.

Deployed appropriately, foresight and hindsight are among the lowest-risk and most rewarding of all knowledge activities. They require minimal capital investment, they are flexible, and each instance brings an immediate return, which can be amplified by linking it with other knowledge activities. Like mentoring, they deserve to be much more widely used.

Foresight and hindsight in the case studies

Feilden Clegg Bradley

FCBa carried out hindsight reviews on the Westfield Student Village at Queen Mary, University of London – the largest student campus in London – at the end of the first two phases of the project in 2004 and 2005. The first was attended by 14 people, including the whole FCBa design team and key people from the QMUL projects and accommodation management teams, the structural and building services engineers, and the quantity surveyors. Participants were first asked to spend 20 minutes reflecting individually on their experience in the project and making notes on Post-Its of the lessons and problems that seemed most important to them. These were then discussed by the whole group for a further 2–3 hours.

After the workshop, the project architect reviewed the Post-Its and her own notes of the discussion, and distilled the lessons and issues onto two A3 pages and a short list of key actions for each party. Everyone was convinced that the event had been valuable, and several aspects of the contract framework, programming and design for phase 3 were changed in the light of the lessons learned.

Twenty people took part in the second review, this time including representatives from the contractor for phases 1 and 2. Post-Its were abandoned as too difficult to see, and were replaced by slides of key issues prepared by the facilitator. Twenty people proved to be too large a number for free discussion, but participants nevertheless considered the event to be as successful as the first. Bringing the contractors into the review process introduced a new dimension, and led to new insights for many – particularly members of the design team. Both FCBa and the client plan to use hindsight again in future.

Arup

Arup has used a different approach to extracting lessons from practice. As an alternative to reviewing projects individually and extracting lessons on whatever topics arise, their skills networks (Arup's name for communities of practice) organised workshops on single topics, where people were encouraged to share lessons they had learned from any of their projects. Key conclusions were then distilled out and published in the networks' pages on the intranet.

Amicus Group, BAA, BP/Bovis Global Alliance, Buro Happold and Lattice Property

There are five further case studies of foresight and hindsight in Chapter 25.

Chapter Eleven
Communities of Practice

The intellectual and emotional stimulus of working in groups that is so valuable in learning can be exploited to help disseminate knowledge, too. Communities of practice (CoPs) – semi-formal, self-governing networks of people who share a common interest in a specific aspect of practice – have emerged in the past few years as one of the most researched and widely promoted techniques for both creating and sharing knowledge.

Professional services firms today are highly task focused and time pressured. Staff have much less opportunity to pursue personal interests and network with professional peers inside and outside the organisation than they had 20 years ago. Short-term operational efficiency may have benefited, but an important mechanism for exploring new ideas and developing new capabilities has been lost. Professional institutions and conferences continue to offer some opportunities, but the proportion of people able to take advantage of them is typically small. Often, contact with peers outside people's immediate environment is limited to events that help meet the institutions' requirements for continuing professional development. Professional services in construction have extra handicaps, working as they do largely in short-life teams that change from project to project, often divided (even in modest-sized practices) between two or more offices, and sometimes with staff away on site for weeks or months. All this tends to inhibit both the development of new ideas and capabilities and the informal sharing of knowledge across and between practices, and to leave people with special interests professionally isolated.

Encouraging enthusiasts

Matrix management was developed to resolve a similar tension between professional (and organisational) identity and the operational requirements of multidisciplinary work. In the matrix organisation everybody has two formal reporting lines, one typically to a business unit or a professional group, and the other to a project team. But matrix structures have high overheads, and they can create as many new tensions and problems as they solve; they do little for the people working in them. Communities of practice can serve the same end, and they are more compatible with a modern professional services environment.

CoPs give people stable intellectual 'homes' that support knowledge, competence and innovation with minimal overheads and without competing with delivery-focused management, and they can adapt organically and flexibly to the individual's and the organisation's changing interests and needs. New CoPs can form as new issues emerge (nanotechnology or renewable energy, perhaps), and old ones close when they become common knowledge, without disrupting formal organisational structures. People can join or leave CoPs – one or several – as their professional interests change, and participate as learners or experts, occasionally or often. And CoPs harness enthusiasm and foster networks and trust in a way that formal structures never do: a volunteer is worth several pressed men. They help to fill the gaps between personal networks, formal groups set up to carry out specific organisational tasks, and the professional institutions.[1]

From the organisation's point of view, CoPs can:

- *Improve business performance*, by giving people quick answers to questions, creating arenas for problem-solving, bringing multiple perspectives into problem-solving and decision-making (and so producing better solutions), strengthening QA, and facilitating coordination and synergy across teams and offices.
- *Develop organisational capability*, by guarding professional standards, promoting mutual understanding and shared values, building trust, developing corporate capability and shareable knowledge assets, exploring emerging issues and providing seedbeds for innovation, connecting the organisation to external networks, and making its commitment to knowledge development visible to clients.

From the community members' point of view, CoPs can:

- *Improve the experience of work*, giving access to expertise and confidence in solutions, helping meet professional challenges, and providing contacts and social structures.
- *Foster professional development*, by providing forums for developing personal skills and expertise, networks for keeping abreast of technical developments, and opportunities to contribute visibly to the organisation and enhance professional reputation.

---

[1] Professional institutions are essentially communities of practice that have become big and formal: the Institution of Civil Engineers, for example, began as a group of six young engineers meeting in taverns and coffee houses.

The most successful CoPs are driven by the personal enthusiasm of their members and the professional and social benefits of participation; corporate benefits are incidental. The benefits to those with special expertise and interests, and to junior staff and new joiners, are different, but they can be equally valuable. And though the psychological commitment implied in joining encourages active participation, most of the work is typically done by an activist minority, while other members simply use the network, learn, and contribute occasionally. CoPs can benefit non-members, too, even if they do not formally publish knowledge resources. Research has shown that they can be important sources of knowledge for people with a passing interest who simply dip into documents and forum discussions on CoP web sites, without joining (though that need not preclude a community's having private space, too). Diverse membership and participation strengthen a community. Experts are inevitably the main contributors of knowledge, but novice members of a community can make an equally valuable contribution by helping the experts to understand the grass-roots perspective and produce more pertinent and helpful publications.

Personal interest is, of course, not enough to enable a CoP to develop. It also needs to be sanctioned by management and given access to some modest resources, and that requires a clear connection to the organisation's corporate aims and aspirations. Usually, the less management is involved the better. CoPs called into being by corporate diktat, or subject to conventional management, rarely thrive.

There is good case study evidence that communities of practice can work well in organisations with hundreds or thousands of professional staff. Little work has been done, though, to study their success in smaller organisations. Clearly, they have no role in a one-man band, and probably none in a practice of ten. Their *raison d'être* is to serve special, and so inevitably minority, interests that are not central to the organisation's routine operations. Only a minority of potential members become active participants; and a critical mass of participants is needed to make them work. All these factors will vary with circumstances, but the indications are that organisations with fewer than 1–200 staff are unlikely to prove fertile soil. For the moment, then, CoPs should probably be considered an experimental option for small practices.

Creating communities

Typical communities of practice:

- Exist outside the formal organisational structure, but are recognised and empowered by the organisation
- Cut across organisational boundaries

- Have a diverse membership
- Are self-organising
- Have voluntary membership and participation, open to everyone
- Inhabit virtual space for most purposes, communicating through phones, email, a community web site and discussion forum
- Meet physically from time to time – important cement for the social fabric.

Many CoPs exist principally to provide a supportive professional network for their members, but they may also act as quasi-mentors for others and become guardians of professional expertise and important originators of knowledge assets. The firm's only contribution is to encourage their emergence and support them with facilities such as meeting rooms, intranet space and IT support, and sometimes a modest time budget for a coordinator.

Communities need a degree of housekeeping. They need a default point of contact for new members and for people seeking answers to urgent questions (a human expertise directory), someone to keep web pages up to date and interface with IT staff to fix problems and make improvements, a moderator in the discussion forum, and some enthusiasts prepared to encourage and help organise collaborative activities and physical meetings. In many CoPs, these tasks are all taken on by rotating coordinators who each volunteer for a limited period, but they can be shared out. A community may also have a chairman who acts as its professional leader, and a senior management sponsor who symbolises the practice's support, ensures that the community gets the resources it needs to function effectively, and helps keep it connected to business realities. Some organisations have found it helpful for communities to have terms of reference that spell out, in simple terms, their technical domain, their objectives, and the basic elements of their modus operandi.

New CoPs may come into being spontaneously in an organisation where others already exist and the concept is familiar, but the first one or two at least are likely to need a positive lead from management. They appear to succeed best when they are built on a nucleus of existing interest, networking and expertise, so the first step in creating a community is to identify this. IBM has developed an 'organisational network analysis' technique – a form of social network analysis – to reveal the people who are regarded as (helpful) experts, and how knowledge flows. This involves asking a reasonable sample of people (IBM suggests 50–100) some simple questions about a specific domain, such as 'Who do you go to for expert advice on X?', 'Who do you talk to normally?', and 'Who do you telephone on Friday night when you have a big problem with X?' The answers can be plotted as a spider's web of connections, which shows who is interested in X and, usually, who the recognised experts are, and to whom people turn as knowledge brokers – typically the people with the best networks.

When a domain of real interest, and the potential members and leaders of a community, have been identified, management can start the ball rolling by discussing the prospects for a CoP with the key people, making sure they understand the CoP concept, giving them start-up time budgets, offering physical and IT facilities, advertising the proposal throughout the practice, and making their support clear. A kick-off event over lunch, perhaps at an external venue, and an intellectual challenge such as a request to develop some knowledge assets, can help. If there is enough enthusiasm to go ahead, management can retreat to a role of visible but quiet support.

CoPs work because they go with the grain of people's natural inclinations: they are essentially natural networks, fertilised by encouragement and modest resources. There are some pitfalls – they may simply fail to thrive, and a few become exclusive, imperialistic and reactionary – but the evidence suggests that they are a good risk: if they do fail, little is lost. They are certainly worth serious consideration as part of any knowledge strategy.

Communities of practice in the case studies

Arup skills networks

Consulting engineers Arup have well-developed communities of practice, which they call 'skills networks'. Arup meet all the criteria to make a success of them: they are a large and physically dispersed organisation (with over 7000 staff and more than 70 offices worldwide), and they have several distinctive areas of professional activity, and a culture that encourages innovation and personal initiative. There are several fields and topics with enough staff to provide the critical mass of members needed to keep a community alive. Despite all the favourable circumstances and a record of success, though, Arup have found that continuing effort is needed to keep communities fresh, lively, and productive. In the past few years they have been exploring ways to do this, and to make the skills networks more supportive of the business.

Two methods that have emerged as particularly helpful are facilitated workshops and the use of stories. Arup have found stories valuable as an informal way to record and spread knowledge, particularly where there is no consensus on best practice, intangibles and context are central, and formal technical guidance is inappropriate – in handling contract disputes and in architects' expectations of visual concrete, for example. Facilitated workshops have proved to be an ideal way to develop and share stories and extract key messages, as well as to support the development of more formal knowledge resources.

Chapter Twelve
Organisational Memory

The indispensability of the written word

Most of the knowledge-sharing that is encouraged by knowledge-friendly offices, networking tools, mentoring, foresight, hindsight and communities of practice is face to face. But face-to-face interaction, despite its virtues, is not a panacea. There are real advantages in recording knowledge in writing, photos, drawings, video and other media – 'codifying' it, in knowledge management jargon – to the extent that this can be done. Codification is the only way to:

- Make knowledge independent of the uncertain availability and fallible memory of individual people
- Detach physical examples and demonstrations from constraints of place and time
- Assemble information that is too extensive or complex to be held in one person's head or communicated by word of mouth or personal demonstration
- Make knowledge accessible to everyone, anywhere, quickly and at any time, even after its original possessors leave – in other words, to create a truly *organisational* memory.

This makes it an indispensable part of a balanced knowledge management strategy.

Recording knowledge sounds an obvious and simple thing to do, but in fact most organisations struggle to do it effectively. Knowledge audits in design practices show that, even when teams carry out post-project reviews (not often!), lessons learned are rarely recorded in a way that makes them readily accessible. A straw poll of the practices involved in one of my research projects revealed that in two-thirds of them the few review reports that were written were 'hardly read at all', and the remainder were only 'read by a few'. Many of the lessons learned in projects are simply forgotten, and would-be knowledge resources of all kinds often languish unread in obscure corners of intranets. This is partly a result of time pressure and the absence of a habit of learning, but not entirely. Tacitness, and the special demands of just-in-time use, make it surprisingly difficult to record expert professional knowledge and lessons from everyday experience in a way that makes them genuinely useful and accessible.

Detailed technical knowledge has always been documented in one way or another. Even among reading-averse architects, textbooks have been a mainstay of professional training since Palladio published his *I Quattro Libri dell'Architettura* in 1570, and there is no real alternative to the written word, drawings, graphs, tables or some other form of documentation for the kind of complex, highly structured, quantitative information that appears in industry bibles and ISO standards. Documenting knowledge such as this is a well-understood process, and it works. It is too labour intensive to be a good model for practices to use routinely, but since relatively little in-house material requires such sophisticated treatment this is only a secondary difficulty. A bigger obstacle is that there are no comparably well-established ways to codify the less formal, semi-tacit knowledge that comes from hindsight reviews, let alone expert knowledge or the fragmentary information that everyone accumulates every day. Much of it can be shared fairly easily in conversation – which at first sight is only words – but the informalities of conversation translate badly into written text, as anyone who has ever read a verbatim transcript will know. Even information written down in personal notes and project records frequently means little without contextual knowledge and mental models that exist only in the writers' heads.

The best that many professional practices in the construction industry can show is a collection of miscellaneous, separate documents, written independently, many of them for transactional and short-term operational reasons, and with little thought given to any subsequent use. Few of these have much value as knowledge resources. Even documents consciously written as knowledge resources rarely get wide use. Would-be users find it too difficult and time-consuming to discover what is there, or to extract, interpret, collate and reconcile incomplete and sometimes contradictory information from multiple documents. It is not enough for knowledge to be recorded; it has to be recorded in an appropriate form, and stored in a way that makes it quickly and easily accessible when users want it.

'Intelligent' search software has been developed in the past few years that attempts to understand meaning and context rather than look mechanistically for individual words and phrases in documents. One of the best-known of these is Autonomy, which also has other impressive capabilities such as analysing the words in a document a user is writing and offering a continuously updated list of relevant material from the corporate intranet. However, even systems like this cannot make bricks without straw. They work well only when there is good material to be found; project correspondence, specs and the other contents of typical construction project files give them little to work with. This, and high cost, makes them attractive only in large practices that have a considerable stock of knowledge-rich documents. When more practices develop resources like this, and prices come down, they may become more widely useful.

Some practices email new guidance notes and similar documents to staff unsolicited, instead of expecting people to look for them – a 'push' rather than a 'pull' approach. Knowledge audits show that this can work for topics that recipients see as directly relevant to their own work. In one practice, for example, a third of staff who had worked on sustainable building projects said they read 'most' of the papers circulated by the sustainability community of practice, and a quarter kept 'most' of them for future reference. It works less well for topics not seen as immediately relevant. In the same practice, only one in eight of the recipients who had *not* worked on sustainable buildings read most of the papers, and hardly any saved them. In general, push approaches have only limited value. Sustainability is an unusually favourable case: it is an emerging topic that many designers find interesting and most feel a need to learn about. Few others have that advantage. And push rapidly becomes counterproductive when volumes build up, as they need to if they are to meet more than a small fraction of information needs. It is usually better simply to highlight additions to the practice knowledge base in a weekly email or on workstation log-in screens, and leave users to click on links that interest them.

Unhappy experience with unused guidance notes and failed initiatives, whether based on push or on pull, has left many professional practices disinclined to invest in recording knowledge in-house. Some are inhibited by the perceived cost, too. They know that it takes time and concentrated effort to formulate one's own thoughts and shape them into lucid text, and even more to elicit other people's and reconcile ideas from a variety of sources. Busy professionals cannot realistically be expected to write documents of more than a few paragraphs when they are already overburdened with their normal work, and to make it possible that has to be cut back. Profit-sharing managers find that a difficult bullet to bite.

But the alternatives are not cost-free, either; their cost is simply less visible. Even when people know who to turn to for answers to their questions, exclusive reliance on one-to-one knowledge-sharing has both real and opportunity costs when people have to repeat explanations to a succession of enquiries, when they are not available when their knowledge is needed, or when they leave the practice and take their knowledge with them. The costs increase with the size and geographic dispersion of the practice and the business value of the knowledge, and they are particularly high when there is rapid growth or staff turnover. Surveys show that staff in design practices where knowledge management is undeveloped typically estimate that they spend 15–20% of their time reinventing wheels and doing rework to correct mistakes and misunderstandings, and as much again looking for information. That is equivalent to 30–40% of staff doing no productive work at all – an enormous and very real cost. The opportunity cost of failing to increase quality, reduce risk, improve marketing or achieve other performance improvements that good knowledge management makes possible are unquantifiable, but they must often be substantial, too. The construction industry in the UK is booming as I write this, and it is not unknown for two-thirds of the staff in an architectural practice to have been with them for less than three years. In a situation like that, one-to-one knowledge-sharing on its own is barely adequate to maintain professional standards and a coherent corporate culture, let alone support effective organisational learning.

Deciding what to record, and how

Codification, of course, is no more a universal solution than sharing knowledge one-to-one. The more tacit the knowledge, the harder it is to communicate it effectively in words and pictures, and even when this is possible, it may be too expensive to be worthwhile. The challenge for managers is to develop approaches to recording internal knowledge that deliver real value in their organisational context, at a cost that makes them worthwhile. Managers need to decide on their business aims for knowledge management, and work out what these imply for knowledge sources and user needs, and for the allocation of resources. The time of experienced staff is always in particularly short supply, and it needs to be divided carefully between one-to-one knowledge-sharing (in CoPs and mentoring, for example) and creating written resources that everyone can use.

In their classic paper *What's your strategy for managing knowledge?*, Hansen, Nohria and Tierney suggest that the overall balance between face-to-face knowledge-sharing and the use of recorded knowledge should depend on business strategy. They found that the leading management consultancies tend to favour one or other of just two business strategies, and to tailor their approach to knowledge-sharing accordingly:

- An emphasis on high-profit business based on offering 'creative, analytically rigorous advice on high-level problems' and bespoke solutions, and employing highly experienced staff – in Treacy and Wiersema's terms, a strategy of product leadership. Firms like this focus their knowledge management efforts on expert mentoring, developing networks and linking people; their systems are designed to facilitate high-level learning, conversations and the exchange of tacit knowledge, and they spend only 'moderately' on IT.

- An emphasis on large overall revenues, based on offering a 'high-quality, reliable and fast' service based on reuse of standard solutions, and a high ratio of junior to senior staff – essentially Treacy and Wiersema's operational excellence strategy. This strategy leads to a focus on developing knowledge bases of codified, reusable knowledge and heavy investment in IT systems to store, disseminate and make it readily accessible.

Both these strategies can be seen in professional practices in construction, and some have followed paths recognisably similar to the management consultants. Buro Happold has gone so far as to base its knowledge strategy entirely on personal contact. In the managing director's words:

> We had the option of a codified knowledge directory, where hard information is placed on a central repository for everyone to access, but in Buro Happold people need to meet, talk and discuss. The hard information will flow from those initial discussions.

It remains to be seen how well this will work. Most organisations opt for a mixed strategy, developing systems and procedures to support both networking and written knowledge in proportions that suit their business model and their culture.

Whatever strategic role is accorded to codified knowledge, selectivity is essential: choosing *what* knowledge to record and *how* to record it according to its value and its nature. Unthinking codification is invariably wasteful and ineffective: the result is a mass of material so low grade on average that it is not worth using. The costs of recording knowledge can easily outweigh the benefits if there are not clearly understood ways to decide, case by case, whether it is worth the effort. The basic principles for this are simple. Knowledge is most worth documenting – writing it down for the first time, or assembling written and tacit fragments into a coherent form – when:

- It has a clear purpose, such as helping people to do a better job, be more efficient, or avoid risks, and either
- Many more people are likely to find it useful than actually possess it, or
- Those who possess it are about to leave the organisation.

There are more factors to consider in choosing *how* to record knowledge, including the context in which people are likely to want to use it, what they already know, and even their psychological make-up: research has shown, for example, that architects, engineers and physicists absorb information most readily in quite different forms. We will return to some of these issues later.

In the remainder of this chapter, we will first discuss the main sources of corporate knowledge and the issues that arise in capturing it from each of them, then turn to the process of codifying it, and finally consider software frameworks for storing it and making it widely and easily accessible.

Capturing knowledge

Existing documents and data

The mass of project documents that exist in every professional services organisation *should* be a knowledge asset, if only a limited one. But when professionals write things down in the ordinary course of work they usually do so with their own purposes in mind, and as an appendage to all their other knowledge. The value of isolated working documents is limited, even if they are shared over an intranet – especially when volumes build up and searches start to produce forbidding numbers of hits. It can be increased by identifying situations in which codified knowledge would be useful to other people, imagining what they would need to know, and getting experts to link fragments together, reinterpret and distil them when necessary, and flesh out the story with their own knowledge to create coherent, shareable, actionable knowledge assets.

The use of operational records in this way is a familiar technique in other fields. Medical researchers, for example, use analysis of medical records to assess such things as the efficacy and side effects of drugs, car manufacturers use warranty claim records to help improve reliability, and post-hoc documentation of responses to past civil emergencies has been found invaluable in enabling quicker responses to new ones.

In professional practice, it is worth mining project records in a similar (if more superficial) way to prepare the ground for hindsight reviews. More ambitious exercises spanning several related projects can also be worthwhile, to discover common factors behind successes and difficulties that are not apparent in the individual projects. Communities of practice are well placed to carry these out and develop knowledge resources from them: they have the expertise and authority, and they can share the work between members to make it less of a chore. However, exploiting existing records in this way is labour intensive, so in practice it is an economic option only in special cases.

Retail businesses that process thousands of similar transactions every day extract valuable knowledge from the transaction records themselves. Analysing who bought what and when can reveal a wealth of information that can be used to tempt customers to buy more: Amazon's 'customers who bought this item also bought' lists are one of the results. On a more modest scale, 'data mining' like this can also be useful in professional practices. Logs of knowledge transactions such as email, intranet searches, file accesses and help-desk requests reveal what knowledge people look for, what they find, and who knows what. This can provide rich insights for knowledge managers, showing (for example) which documents are valuable and which simply clutter up the system, where there are gaps, and from whom it would be most useful to capture tacit knowledge. Automated email analysis calls for specialised software, which is worth considering only in the largest organisations, but some other transactions can be analysed effectively and at much lower cost by hand. Logs of help-desk requests and intranet searches, for example, can be used to help manual compilation of frequently asked questions, which are a good way to make key information more accessible. Some organisations already do this.

In practice the most valuable existing material may be guidance notes and other documents that were written as knowledge resources, but which failed to make an impact because they were in an inappropriate form, or simply inaccessible. Sometimes their utility can be transformed by fairly simple restructuring and remounting in a better software framework, and at the least they provide useful starting points for developing a new, better-targeted content. Whatever its quality, it is usually easier to recognise the strengths and weaknesses of existing material and build on that than to start from scratch.

Personal knowledge

As we saw in an earlier chapter, it is inherently difficult for people to articulate personal expertise in a way that allows it to be shared electronically, even when they can communicate it successfully face to face. Sometimes it is simply impossible. It is a particularly daunting task when the topic is broad and the expertise is high level, as it often is in the situations where organisations typically want to capture personal expertise: to make their experts' knowledge more readily accessible in a knowledge base, and when the Mr Inghams of the *New Yorker* cartoon that we met in Chapter 1 retire. The dream of reducing expert knowledge to sequences of logical decisions and encapsulating it in 'expert system' software, which seemed to promise so much in the 1960s, has been realised only in very narrow domains.

Whatever the situation, a systematic approach helps: thoughtful selection of the knowledge to be documented, to minimise the task; careful consideration of users' likely knowledge and what they need to be told; logical structure; and forward planning to ensure that there is enough time. When the occasion is departure for another job, retirement or redundancy, it is too late to begin capturing more than the simplest knowledge a few weeks before people leave. They will invariably be too busy clearing up, and they may be losing interest, or (in the case of redundancy) resentful. Sharing knowledge at that time can seem like an unwelcome chore, or even giving away a personal asset for no return. Whenever possible, therefore, the process should be spread over months, or even years. This can be done by involving all experienced staff in mentoring, foresight and hindsight reviews, communities of practice, writing for the knowledge base and other knowledge-sharing activities as a matter of routine. Then, by the time they leave, much of their knowledge will already have been passed on, and key parts of it will have been documented.

Research on expert systems gave rise to several formal methodologies for capturing ('eliciting') knowledge, but these have only limited value in a professional practice context. Informal 'brain dumps' are little better; they can capture only small amounts of knowledge, and they are unlikely to be usable as just-in-time resources as they stand. When speed is unavoidable, face-to-face, question-and-answer knowledge transfer probably provides as good a model as any to follow. The starting point – the question – becomes in this context a conversation between the expert and colleagues (and, when the expert is leaving, with his successor) to compile a list of specific areas of expertise that are most likely to have future value, and specific aspects of them that need to be documented. This can be circulated among colleagues to spark additional ideas. With questions compiled in this way as a guide, it becomes much easier for the expert to produce well-focused and helpful material in notes, or in interviews with an intermediary. These may still not be directly usable, but they should provide good raw material that can be edited into an appropriate form.

MIT's learning history process is similar in some respects, and offers further hints. MIT found that it was helpful to have two interviewers (ideally with complementary skills in the technical domain and facilitation) in order to leave one free to reflect and make notes while the other interacted. Their use of verbatim quotations alongside a distilled interpretation can also be worth copying.

Lessons from practice

It makes sense to exploit existing records more fully, and to capture experts' know-how as a knowledge resource, but these are only the tip of the knowledge iceberg. Much more accumulates, unrecorded, in people's heads day by day, and the big prize is to capture this – or even a fraction of it.

Most of the documented knowledge in the public arena – books, published reports, manuals and so on – originates in formal research of one kind or another carried out variously by universities, industry research departments, professional institutions, journalists, independent authors and others. All this activity has the explicit objective of finding things out or collating and reinterpreting existing knowledge, and then communicating the results to readers. The knowledge that arises in the course of experience is quite different: it is simply an accidental by-product of an activity that has entirely different objectives. It can arise anywhere and at any time, unpredictably; it is typically fragmentary – one or a few pieces of a larger knowledge jigsaw rather than a complete section; and writing is rarely a core professional skill for the people who have had the experience and learned lessons from it.

Today, as in the past, the task of documenting important information and guidance is seen in most organisations as the preserve of senior professionals and managers, technical experts and specialist support staff. This is reasonable when work processes are developed at the top, and the majority of staff do stereotyped jobs in which they are expected to follow the rules. It has less value in professional services organisations, where expertise and opportunities to learn are widely spread, and staff work with a high degree of autonomy. Little of the learning at the grass roots – which makes up the majority of the learning in the organisation – ever reaches the ears of senior staff, and even if it did they would not have time to digest and document it. Dedicated support staff can have a role, but the cost of using them to trawl continuously for new lessons would be prohibitive. Where they *are* used – in Bovis's *i*konnect system, for example – capturing lessons learned is typically ancillary to a wider role as information brokers, and a great deal still passes unrecorded. The only practical solutions are:

- To give more, and ultimately all, staff the responsibility, encouragement, opportunity and tools to allow them to record lessons learned as they arise
- To take advantage of other activities such as hindsight that offer particularly rich opportunities to learn.

Hindsight reviews – and to a lesser extent foresight reviews – are ideal opportunities to capture and codify knowledge. All the basic steps are inherent in the occasion: identifying exactly the kind of notable events and issues that are worth documenting; gathering illustrative material; understanding causes; working out ways to do better; articulating all this; and adding personal anecdotes that help bring the story to life. This was discussed in Chapter 10, and we need not revisit it here. The larger challenge is to capture the lessons that arise in everyday practice, and we will return to that shortly.

Documenting knowledge

Lessons learned from practice and distilled from outside sources can be recorded in several ways:

- Embodied in formal procedures
- In guidance, examples and case studies published as individual documents on the practice intranet, or circulated by email or in hard copy
- In a web-like knowledge base, where content can readily be searched and navigated, cross-referenced with hyperlinks, and viewed directly without opening separate documents
- In stories told in seminars or newsletters
- As video or audio recordings, for example of demonstrations or talks.

Several of these are widely used, if not often as part of a process systematically designed to add to the organisational memory and develop staff expertise. Formal procedures need no further comment here; they have other objectives, and any knowledge management benefits are incidental. Collections of separate documents demand too much of users' limited patience to be effective in practice; they look set to be replaced for most purposes by web-like knowledge bases. The increasing speed of data networks promises to make audio and video recordings more practicable and widely used than they have been in the past, probably embedded in knowledge base pages; in some cases, pictures really can be worth a thousand words. Stories used to be dismissed as a serious medium for communication, but they have received increasing attention in the past few years. For some purposes they can be the most compelling format of all.

Whatever form knowledge resources take, there are some general principles that apply. In the remainder of this chapter we will look first at some of the most important of these, then at the role of stories, at the organisation of knowledge bases, and finally at the IT options for giving practitioners rapid access to just-in-time knowledge.

Just-in-case versus just-in-time

Just-in-time knowledge – the knowledge people need to get them over an immediate project problem and move on – is particularly important in professional practices. It is this kind of knowledge – rather than the just-in-case knowledge that people get at university, or when they read a book or a journal on the train – that does most to increase productivity, improve quality and reduce risk. One of the commonest mistakes in documenting knowledge is failure to consider whether it is being recorded for use just-in-case or just-in-time.

The unthinking default is to write in the just-in-case style familiar from books and journals. This discusses a complete topic (if only a narrow one), deals principally in general principles, and expects readers to begin at the beginning and carry on to the end. It is ideal for introducing readers to whole areas of knowledge, but it has limited value for the main purposes of knowledge management. Practitioners usually need only one or two pieces of specific information to complete a knowledge jigsaw and solve their problem, and they want them in a hurry, just-in-time. Information like this is better provided piece by piece than buried in a larger picture, and this calls for a different way of organising and presenting it. Most textbooks and journal articles are at the just-in-case end of the spectrum. Technical manuals and encyclopedias, with their greater subdivision, more detailed indexes and (in the case of encyclopedias) alphabetic arrangement lean more towards the just-in-time. Electronic resources such as Wikipedia, with additional tools such as search engines and hyperlinks to give even more selective and rapid access, are the nearest we have yet achieved to the ideal of instant knowledge on demand.

There is still a place for just-in-case material. It can, for example, be well worth developing slide sets, videos and stand-alone documents for training courses that will have to be repeated several times, and for new entrant induction. There are good reasons to produce user guides for software systems, technical manuals detailing CAD standards, and procedure manuals for HR and financial management. House magazines and newsletters can be useful vehicles for keeping people in touch with an organisation's work and staff, and with notable events in the wider world, and for introducing them to new developments in professional practice. However, few people will want to refer to documents such as these more than occasionally, except perhaps for a period after joining a practice or taking on a new role. What practitioners most need are resources that can provide answers just-in-time to the new questions that arise unpredictably every day, and that is our main concern in this chapter.

Completing jigsaws

Gabriel Szulanski, Professor of Strategy at leading business school INSEAD, carried out a landmark study of knowledge-sharing based on a highly detailed study of 12 American firms in the mid 1990s. He found that many of them struggled to share best practice and other knowledge, and identified four main barriers. In decreasing order of importance, these are:

- The prior level of related knowledge, which he called *absorptive capacity*. In Szulanski's words, 'A recipient that lacks absorptive capacity will be less likely to recognize the value of new knowledge, less likely to re-create that knowledge and less likely to apply it successfully.'
- Poor understanding and explanation of the reasons why a practice worked in its original context: Szulanski called this *causal ambiguity*. He found that people often fail to identify the factors that are crucial to success, typically because they see everything through the lens of their own expertise; technical experts, for example, tend to ignore human factors. Even when they do understand the success factors,

authors may exclude some from their explanation because they think they are politically unacceptable.

- The pre-existing relationship between source and recipient. Szulanski found that knowledge transfer often fails because a writer does not understand readers' circumstances in enough detail to know what they will find useful, and what they will need to have explained; a close working relationship in the past makes communication much easier. Szulanski called the lack of this background of shared experience, common language and assumptions *arduous relationship*.
- *Recipient motivation* – which most managers expected to be the most important barrier – did prove to be a factor, but a much less important one than the others.

Just-in-time knowledge resources need to be written with the top three of Szulanski's barriers clearly in mind, and especially the first, which he found to be by far the most important.

The significance of 'absorptive capacity' (and of 'arduous relationship') is, of course, entirely predictable from the jigsaw model of knowledge that we met in Chapter 2. Information – the stuff we can codify – is equivalent to knowledge only when it fills gaps in users' knowledge jigsaws and enables them to do things that, without it, they could not. It is impossible to provide the information needed to do that without understanding what people's gaps, and the picture around them, are like. Authors of documents that aim to share knowledge therefore need to think carefully about what prior knowledge their readers can be assumed to have, and about the context in which they work. It is a frequent criticism of published case studies in construction, for example, that they lack the detail and contextual information that readers need in order to put them to practical use.

Knowledge audits have shown that one of the commonest reasons for technical information and guidance failing to influence practice is that its authors have forgotten how little they knew when they were juniors, and fail to imagine successfully what their readers will need to know. The opposite failing is also common: inability to resist the urge to include prefatory explanations that are superfluous and serve only to obscure the important material. Both of these, and the other risks that Szulanski identified, need to be kept in check by authorial vigilance and constructive comment from representative members of the intended audience.

Telling stories

With a clear focus on the reader's point of view, and good writing style and document design, most technical knowledge can be communicated effectively in the impersonal, topic and logic-based structures familiar from technical notes, journal articles, textbooks and web pages. But the conventional forms are not good at everything.

When the business value of knowledge started to enter the corporate zeitgeist in the 1990s the World Bank asked one of its managers, Steve Denning, to look into the possibilities of improving its information-sharing. He became enthused, and started a campaign to persuade his colleagues. But his logical arguments and charts had little effect. The Bank had seen its mission of reducing global poverty and improving living standards exclusively in financial terms for 50 years. Its job was to provide loans and grants, and the reaction to Denning's presentations was uncomprehending: 'Knowledge? We're a *bank*.' And then, in casual conversation, he heard a story. In June 1995 a health worker in Zambia logged onto the Centers for Disease Control website in Atlanta and got the answer to a question on how to treat malaria. Denning started to use the story to show that knowledge could be as valuable a resource as money, and found that it sparked interest in a way none of his rational arguments did. Attitudes changed, and sharing knowledge became a major plank in the World Bank's strategy.

Storytelling has been emerging slowly as a topic for research and a tool for business for 30 years, and interest has snowballed since the late 1990s. It is in danger of being oversold at the moment, but there is no doubt that it can be a valuable technique, as Steve Denning's Zambian example shows.

We all use stories every day in casual conversation, even in a professional context: they are the most natural vehicle for explaining many of our ideas, and they are understandable and memorable in a way that concise bullet points are not. They work well for both teller and listener. As Cornell professor Robert Frank has put it, 'If you can wrap a story round an idea, it seems to slide into the brain without any effort.' What has changed recently is an appreciation that stories can be equally valuable in the more formal contexts of presentations, teaching material and knowledge bases.

Stories are particularly good at:

- Engaging interest
- Igniting action
- Sparking imagination and creativity
- Making high-level ideas and abstract concepts more meaningful
- Sharing knowledge in which context is crucial
- Promoting norms, values and culture change
- Evoking emotion
- Increasing confidence.

They are also unusually memorable: many of our personal memories are encapsulated in stories, and storytellers in preliterate societies are credited with remembering histories and myths of amazing length.

The managing director of management consultants Arthur D. Little has suggested that stories 'may prove to be the single most powerful technique in business organizations where personal choice must be the centerpiece in making change happen.' They inspire, and they help develop competence in action rather than simply knowledge of facts.

The key features of successful stories appear to be that:

- They are about people in situations with which the hearer can identify, what they do and why, so they put action in a personal context.
- They are told from the point of view of a single protagonist.
- One thing leads to the next, and they have a beginning, a middle and an end. (This helps to make them memorable, mimicking the classic trick of memorising a speech by associating each part with places on a familiar route.)
- They include an element of surprise.
- They have a positive ending.
- They evoke vivid mental images.
- They are brief, focusing on the essence of an idea.
- They ring (and ideally are) true.
- Often, they use analogy and metaphor, leaving hearers/ readers to make connections with their own situation. This active engagement makes the implications more memorable than they would be if presented ready-formed.

Storytelling missionaries insist on the importance of crafting stories with great care and polishing them with practice. This may be the ideal, but the evidence suggests that the main elements – a protagonist, a focus on action, brevity – are enough in themselves to make a story communicate more vividly and memorably than impersonal abstractions. If stories had to be as perfect as some suggest in order to work they would not have been so pervasive a feature of human discourse throughout history. The popular media invariably use 'human interest' anecdotes to help explain science and technology; MIT stresses the importance of verbatim quotations in its learning history format; and Harvard Business School bases its MBA teaching almost entirely on discussion of case studies. None of these conforms strictly to the supposed ideal, but they nevertheless realise many of the benefits of the story format.

Stories, then, are potentially important weapons in the knowledge codifier's armoury. They can be particularly helpful in stimulating change, and the prominence of case studies – which are first cousins to stories – in design literature suggests that design practice is a promising area for them.

Organising knowledge

The usability of any work of reference depends first and foremost on how it is organised. When its structure is good, it is intuitively obvious where a particular piece of information is likely to be; when it is not, even excellent content loses much of its value. Occasionally structure is self-selecting (as in dictionaries), but more often there are several, even numerous, possibilities, and that is certainly the case with knowledge bases. The first step in developing them is to design a structure that reflects the way users naturally think and so makes it easy for them to find the content that is most relevant to their needs.

To see the kind of choices that need to be made, consider a knowledge base for an architectural practice. Users may be thinking at different times in terms of building types (offices, hospitals, houses), basic components (foundations, walls, roofs), materials (concrete, steel), subdisciplines (masterplanning, interior design), activities (project management, drawing), legislation and standards, broad themes such as sustainability, or in other terms. They might be looking for information about principles, detailed procedures, successful details, or pitfalls to avoid. If these were all made top-level headings (akin to chapter headings in a book) the result would be confusion: information about insulating the junction between a brick wall and a roof might appear under components (walls), materials (bricks), sustainability (energy efficiency), or in any of several other places. Documented knowledge needs to be organised in such a way that the location of any particular piece of information is self-evident.

This is difficult on the printed page but, fortunately, IT systems offer more options. Hyperlinks make cross-referencing simple, and content can be made to appear in more than one place. There are no significant space constraints on a modern server, and repetition does not jar in a reference system. Content can even be split between separate systems that are organised on different principles to suit their content, but stitched into a seamless whole by links. It is helpful, for example, for users to be able to see the context in which a particular lesson was learned, but it is *unhelpful* to clutter up a topic-based section with project descriptions. This can be avoided by recording topic and project information in separate systems, with cross-references – an approach that also neatly separates the project information that is important in marketing from the topic information that is important in design.

Software frameworks

Expert guidance, information picked up from external sources, and knowledge gained through processes such as hindsight reviews and exit interviews, have all traditionally been recorded in stand-alone documents of one kind or another. But, as we have seen, these have serious limitations, including:

- *Limited visibility and accessibility.* The people who could benefit from them are often unaware that they exist unless they turn up in a document directory or a search hit list. When this is long it may not be clear which sources are valuable, and even after a promising document has been identified it may need a further internal search to find specific information. Searches are also vulnerable to differences in terminology (as users of software 'help' systems are often made painfully aware!), and it is easy to miss useful sources.
- *Fragmentation.* Hindsight reviews and day-to-day experience generate information randomly on a range of topics, often in small scraps. Documents such as workshop minutes, for example, usually include small amounts of information on several topics, and there may be information about any one in several documents. This fragmentation severely limits their value for busy practitioners, few of whom have the patience to search and review multiple documents.
- *Lack of coherence.* Most stand-alone records are written without reference to other material on the topics they cover, so (the few) people who do take the trouble to collate information from multiple documents are likely to find duplication and contradictions. This simply replaces one problem with another.
- *Inappropriate content and presentation.* Lack of key information, inaccurate identification of causes, just-in-case style and other deficiencies can all make written knowledge hard to understand and interpret correctly.

Shortcomings such as these make stand-alone documents unhelpful for busy practitioners. To be useful as an everyday, just-in-time resource, knowledge needs to be recorded in a way that makes it part of a visible, consolidated, coherent, high-quality and well-presented knowledge system where it can be easily found, is easily understood by its audience, and gives users the information they need to act correctly.

Expert review of source material, consolidation and careful authoring can overcome all the difficulties of stand-alone documents apart from visibility and accessibility, and of course that is how textbooks and reference documents are produced. But producing reference documents (whether long or short) takes skill and a great deal of time, and in a professional practice it inevitably devolves onto senior staff and technical experts, who are among the most overburdened people in the organisation. As a result, development and maintenance become spasmodic activities, leading to gaps in coverage and long revision cycles. When, as is usual, there is little knowledge flow from the grass roots upwards, a high proportion of potentially valuable experience is simply ignored. Often, content is more a reworking of textbook material with a practice-specific gloss than a systematic attempt to collate accumulated lessons learned. Overall, this approach is too labour intensive and unwieldy to have more than a niche role in businesses where knowledge bases have to cover a wide range of topics, and knowledge is constantly evolving. Today's needs call for a different approach.

The storage and distribution of codified knowledge need rethinking, too. Until personal computers and networks came into widespread use there was little choice: paper was king. Documents were stored in company filing systems, practice libraries and personal caches, and users had to rely on catalogues and indexes of varying quality, on memory ('I'm sure I saw it in a blue book on the second shelf. . .'), and on riffling through pages to find what they wanted. The whole system was slow, uncertain, and expensive in labour and space. The appearance of personal computers in the 1980s did little to change this; they were used to *create* documents, and laser printing engulfed offices in more paper than ever before. With few PCs in a typical practice and even fewer networks they had only limited value for storing or distributing information, but as they spread, their potential to replace paper for communication and record-keeping, and in reference libraries, became obvious.

This has now been achieved to a large extent (though paper has shown a remarkable reluctance to disappear entirely!). Filing cabinets have become rare, offices look tidier, and on the whole communications, product data, standards documents and public information have indeed become easier to find and retrieve. Project documents have, too, in many offices. Other information and knowledge, though, has often become *less* accessible, just when the growing technical complexity of professional knowledge, the emergence of new concerns such as climate change, increasingly demanding clients and an influx of inexperienced staff make it more needed than ever. Replacing paper filing systems with electronic files in electronic folders helps only when it is obvious how documents should be organised – contracts, specs and drawings by projects, email by date and source and so on – and there are relatively few originators and potential users. It fails for information and knowledge that are potentially relevant to almost anyone, in a wide variety of contexts.

The explosive growth of the World Wide Web points the way to the first part of the solution. The reason why the Web has become the world's richest and most widely used knowledge resource is that it enables specific information to be retrieved from anywhere in the world, often within a minute or two, through an interface anyone can use, and with little or no knowledge of where it originated or where it is stored. Three key ideas make that possible, and enabled the Internet – which is simply a public super-network that enables private computer networks to be connected together and exchange data – to become such an effective tool for publishing and accessing information:

- *Hyperlinks*, which enable users to move instantly between related websites and related pages within websites
- *Search engines*, which know the location of every word in almost every page on the Web
- *Common standards* for web pages (HyperText Markup Language, HTML) and for communication between websites (HyperText Transfer Protocol, HTTP).

These are the keys to creating rich and readily accessible knowledge resources for private use inside organisations, too.

The data transfer protocols developed for the Internet were quickly taken up within organisations as the basis for their internal networks.[1] Internal webs, though, have developed more slowly. In its original form, the Web had two characteristics that limited it largely to corporate use and one-way information transfer, from publisher to user: websites and pages could only be created using special-purpose software that is expensive and far from intuitive to use, and maintaining them so that all the hyperlinks continue to work as they grow and change was difficult and labour-intensive. Conventional websites can make information visible and accessible, but they leave the problem of creating and maintaining them untouched. They are simply too demanding in expertise and time for more than limited internal use in most professional practices.

This has all begun to change in recent years with the development of software that makes it almost as easy to build websites and create pages as it is to use them, complete with all the features that have made the Web so successful: not only text but also pictures, hyperlinks, navigation bars, search engines and so on. This, and variants on it, are starting to transform the Web from an essentially one-way channel into a two-way channel, and to create the embryo of what has come to be called the 'participative web' or *Web 2.0*, a medium as much for individuals and collaboration as for organisations and publishing. Blogs (personal diaries published on websites), wikis (websites based on information written and uploaded by their users, most famously Wikipedia), social networking sites such as Facebook and LinkedIn, social bookmarking sites such as Digg and Reddit (which

---

[1] Hence the name 'intranet', often – and wrongly – used to refer to any private network, whether it uses the Internet protocols or not. True intranets are usually accessed through web browsers.

enable people to store, organise and share lists of web pages they have found useful) and the shopping sites that invite customers to post personal opinions of products, are all part of this democratic, interactive, do-it-yourself web.

The ideas and software behind Web 2.0 provide the second part of the solution to the problem of creating truly effective knowledge resources within organisations.

Wikis: webs made by their users

The key Web 2.0 software tool from a knowledge management point of view is the *wiki*, which was developed by software engineers in the mid 1990s for project team collaboration. This has proved to be an almost ideal framework for knowledge bases as well.

Wikis neatly sidestep the main limitations of both stand-alone documents and traditional websites. They are websites in which users – usually but not necessarily *all* users – can edit existing content or add new text, images, pages, internal and external links and attached documents at any time, from their own browsers and with no need for special software or IT expertise. Most wiki software has sophisticated navigation, search and quality control tools built in, and provides flexible ways to organise information and give users rapid access to content.

Wikis allow new information gleaned from external sources, from project reviews and from day-to-day experience to be integrated seamlessly into previously existing material, and to be easily updated. They make an equally good home for carefully authored, authoritative guidance from experts and for the scraps of knowledge that arise anywhere, at any time. They can act as a portal into other information resources. And with good design they can make all their content into a coherent and accessible knowledge resource capable of delivering just-in-time knowledge to anyone familiar with the Web with just a few mouse clicks. Finally, they allow the effort of developing and maintaining content to be distributed between many people, so that it need not be an undue burden to anyone.

These features make it more practicable than ever before to exploit the power and user-friendliness of websites in storing and sharing knowledge inside organisations. Wiki software provides an ideal framework for any information system that needs to draw semi-structured content from numerous people and make it available to others, networking tools, project directories and knowledge bases alike. Most importantly, it makes it feasible for the first time for practices to build up a rich organisational memory, present it in a user-friendly way, and keep it up to date.

In the past few years, wikis have been widely adopted as a framework for knowledge bases across industry and as virtual homes for communities of practice. They are being used, too, for their original purpose as collaboration and general communication tools, where they can replace discussion forums, bulletin boards, and broadcast emails. A few household-name users are listed in a box at the end of this chapter. Wikis are even being used, with edit facilities restricted to a few staff, as the basis for public websites where quick development and avoidance of the need for expertise in HTML and of conventional web development tools are more valuable than sophisticated graphic design and technical features.

Most wiki software shares a number of common characteristics:

- Documents can be *created* as well as read in any web browser. No extra software has to be installed on users' PCs, and no special software skills are needed.
- Simple editing and formatting needs only simple syntax; usually, full HTML can also be used if users wish.
- New pages are created and linked automatically on demand.
- Links to stand-alone files and external websites can be included.
- All changes are automatically signed by their author, stored, and can be reversed.
- Pages are stored as HTML in a database running on a web server, controlled by a CGI (Common Gateway Interface) script.
- Most of the software is open source, so there are no licence fees.
- The software can be changed and extended to tailor capabilities to local requirements.

Beyond the generic features, wiki software packages all have different details, and some are more appropriate for design practice than others. The mock-up in Figure 12.1 illustrates some of the facilities that wikis can provide, including an expandable contents tree, an A to Z page index, tabbed access to the editable version of each page, its history, and a discussion page, and links to other pages in the wiki, other wikis, and external websites. Some of these can be seen working in Wikipedia at www.wikipedia.org.

Many of these features can be found on conventional web and intranet sites. What sets wikis apart is their ease of use, and the facility for any user with appropriate permissions to amend their content and structure. Inevitably, this leads to fears about anarchy and quality. In practice, though, this is a negligible risk. Wiki software includes powerful features to encourage good behaviour and allow damage to be repaired. There is social pressure to be constructive, because changes are traceable to their authors, and they can all be reviewed and reversed (usually with a single mouse-click) at any time. In a professional practice, traceability to authors allows doubtful users to assess the standing of information on exactly the same basis as when they ask a colleague, and light-touch oversight from senior moderators can provide additional reassurance without making undue calls on their time.

For its first six years Wikipedia was open to additions and edits from anyone without prior accreditation, and the oversight provided by other users proved good enough for the highly respected journal *Nature* to conclude that Wikipedia's content on scientific topics was, on average, as good as Encyclopedia Britannica's. Deliberate distortion, vandalism and strong emotions on pages devoted to politics and prominent people have recently prompted the introduction of a systematic checking process, but these are unlikely to be issues in a practice knowledge base. In a practice context, worries about the quality of content are largely misplaced. A tidy and logical site structure can be ensured by launching the system with a basic framework of topic headings and page templates in place, and stylistic standards can be set by seeding it with exemplary contributions.

Another fear is that free, open-source software will not be robust and scalable enough to rely on in a corporate environment. Again, Wikipedia – which uses the open-source MediaWiki package – proves otherwise. Between its launch in 2001 and the time of writing in 2007, the English-language version (there are 252 others!) grew to over 2 million articles, 10 million pages and several million pictures and hyperlinks contributed by over half a million people, and it is still growing at something like 2000 articles a day. There have been over 100 million individual edits, and the site gets tens of millions of hits every day. It appears to suffer from no more technical problems than conventional websites, and it has fewer than most.

Setting up a wiki is technically straightforward, as Feilden Clegg Bradley's experience (described in detail in the FCB case study) demonstrates. One of their junior architects with an interest in IT installed and configured it in a few days. But, as with all the IT tools useful in knowledge management, there is more to a successful wiki than software that works. Its purpose, structure, content, management and relationship to other knowledge management tools and activities all need careful thought.

A wiki can play a variety of roles from pure knowledge base, project directory or networking tool to the foundation for a practice's entire intranet at one extreme and a short-life collaboration tool for a single project at the other – or all of these. Wikis make ideal virtual homes for communities of practice, giving members all the facilities they need to both collaborate privately and to interact with and publish to the practice as a whole. The more uses to which wikis are put in a practice, the more opportunities there are to enhance their value by linking them to each other and to knowledge management activities in general. We have seen how a wiki knowledge base can be made the confluence for knowledge contributed by practice experts, lessons learned in foresight and hindsight reviews, and the fruits of personal

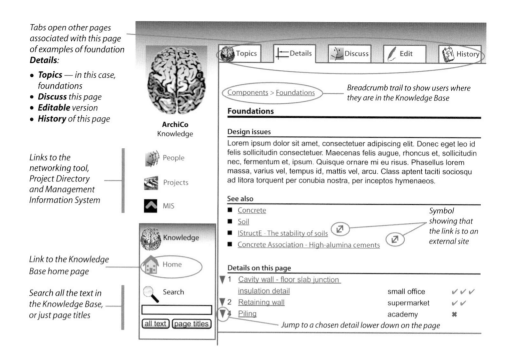

Tabs open other pages associated with this page of examples of foundation **Details:**

- **Topics** — in this case, foundations
- **Discuss** this page
- **Editable** version
- **History** of this page

Links to the networking tool, Project Directory and Management Information System

Link to the Knowledge Base home page

Search all the text in the Knowledge Base, or just page titles

Figure 12.1   A mock-up of a page in a knowledge base.

learning by staff in general. Knowledge bases, project directories and networking tools can all be enriched by hyperlinks between them and to other corporate systems. Links from knowledge base pages to project records can show the context in which lessons were learned, or simply provide illustrative examples, and links to personal pages show users how far they can trust the content. There are natural connections with communities of practice too. CoPs can be important contributors to a knowledge base (as well as publishing through their own wikis), and they are well placed to act as content moderators, taking responsibility for keeping an eye on articles within their areas of interest and rolling back unhelpful contributions. More possibilities become evident as systems develop and people become adept at using them.

Despite their potential, though, wikis (like all knowledge management tools) need to be sold persuasively to staff. Chastening experience quickly disabused knowledge management pioneers of the expectation that, when systems are provided, users will flock to them; they won't. Systems need to be self-evidently useful the first time staff see them, and be launched by senior staff, with demonstrations, opportunities for hands-on trial, and tutors who can explain unfamiliar aspects such as editing facilities. Even after a successful launch people can forget quickly. New knowledge tools need to be kept in people's

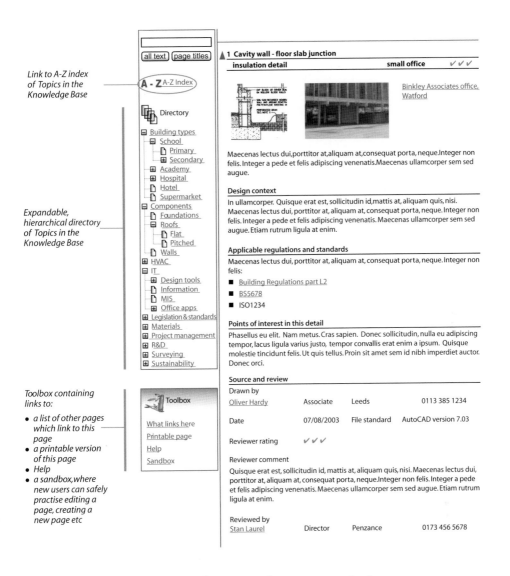

Link to A-Z index of Topics in the Knowledge Base

Expandable, hierarchical directory of Topics in the Knowledge Base

Toolbox containing links to:

- a list of other pages which link to this page
- a printable version of this page
- Help
- a sandbox, where new users can safely practise editing a page, creating a new page etc

---

minds by repeated reminders to try them out, continuing encouragement from managers, and (in the case of knowledge bases and project directories) a flow of attractive new material until access logs show that using them has become a widespread habit. It can take a surprisingly long time for people to develop new habits and replace managerial push by the grass-roots pull needed to keep systems healthy in the long term.

Post-launch, of course, wiki knowledge systems, like any other, need ongoing management and maintenance. Someone – or a group of people – needs to be responsible for introducing new joiners to them, encouraging contributions, making constructive links with other knowledge-sharing activities, setting access permissions, resolving any disputes among contributors, and other maintenance. Wikis also need the technical maintenance usual with any IT system. This is a small price to pay for a corporate memory that actually works.

The key strengths of wiki knowledge bases

- Collaborative development of content, tapping into everyone's learning and spreading the effort.
- Quick, incremental additions, from a single sentence to a major article, all in context.
- Robust content control, with minimal 'big brother' bureaucracy and overhead cost.
- Quick access to information through multiple routes – a topic directory, alphabetic index, full-text search, and multiple cross-linking within the wiki and between it and other systems.
- Encourage contributions, use and ownership from everyone.
- Complement other knowledge management tools and resources – networking tools, project databases, other intranet and web resources, communities of practice and so on.
- Also valuable as a team collaboration tool and general web publishing tool.
- Well proven, cheap to acquire, maintain and use, based on common data standards (so unlikely to become obsolete, and avoid lock-in to single vendor), extensible.
- Also available as a commercially hosted service, allowing organisations to set up a wiki without need to install and maintain software on their own servers.

Notable wiki users

Wikis are already widely used in other industries. Users of one of the most popular open source packages, TWiki, include:

Amazon.com
AT&T
Boeing
CNN
Compaq
Disney
Ericsson
FedEx
Ford Motor Co.
General Motors
IBM
Intel
Lockheed-Martin
Matsushita

Continued

NASA
New York Times
Nokia
Philips
SAP
Siemens
Sun
Texas Instruments
US Government
Xerox

Codified knowledge in the case studies

Feilden Clegg Bradley's wiki knowledge base

FCB's new wiki Knowledge Base is designed to make it easy for everyone to record new knowledge as it arises, and find it quickly when they need it. FCB decided to use wiki software because it suits their traditionally collaborative, inclusive approach to knowledge better than more conventional solutions; it enables them to integrate existing information resources into the new knowledge base; and it is flexible, technically powerful, and highly affordable.

Their wiki is based on the open-source TWiki package. It has six main topic headings – Buildings, Materials, Environment, Practice, IT Technical and FCB Community – each of which is further divided into subtopics (General, Concrete, Masonry and so on in the Materials section, for example), with several individual pages in each subtopic (such as InSitu and PreCast within Concrete).

They have used the same wiki software to create the main home page for the practice intranet. This links to the Skills, Project, Image and Certificate databases and a variety of other resources and administrative tools such as practice procedures, time sheets and external sites as well as to the knowledge base itself.

Edward Cullinan Architects and Aedas

ECA and Aedas have also adopted wiki software for their knowledge bases. In Aedas's case this is a development from software developed in-house for the practice's management information system rather than a special-purpose wiki package, but it provides the same basic functions, including in-browser content creation and editing, and easy hyperlinking.

Storytelling at Arup

Arup have started to use storytelling to meet a perceived need for a less formal vehicle for sharing knowledge, particularly in contentious areas where experiences need to be shared between small groups (contract disputes, for example), and in areas where best practice is still uncertain. They expect it to become a key technique for knowledge-sharing in communities.

Chapter Thirteen
Personal Knowledge Management

Equipment for the mind gym

Until recently, knowledge management has been seen as an essentially corporate concern. This is reasonable, insofar as its main aims are to improve organisational learning and generally make the organisation more than the sum of its individual parts. Many of the underlying ideas, though, apply equally at a personal level. A systematic approach to learning from experience pays dividends for individuals just as much as it does for organisations, and most professionals have personal stores of information that would be more valuable if they could be searched more quickly.

Corporate knowledge tools serve ends such as these only to a limited extent. Most personal learning opportunities fall outside the scope of project reviews, for example, and it is only a partial solution to have a personal electronic library included in a corporate index. Besides, personal needs such as collating information for a presentation or project are often too transient to be met by corporate systems, which are designed to deal with information of long-term value. The tools and techniques for acquiring, creating, storing, organising and accessing knowledge that are most helpful for individuals are often distinctly different from those that suit organisations. Recognition of that has led to the emergence of interest in *personal* knowledge management – tools and techniques that people can use to reinforce their own memory and capabilities.

Like corporate knowledge management, this is a development built on familiar foundations: tools such as paper 'to do' lists, calendars and diaries, address books, Filofaxes and their electronic equivalents such as PDAs and Microsoft Outlook have been a routine part of professionals' working life for years. The term 'personal knowledge management' (or PKM) is normally used to refer to tools and techniques that provide more powerful facilities than these. This is a rapidly developing field, in which there is still a multiplicity of offerings competing for attention and market share. Many of today's IT-based PKM tools may well disappear within a few years, some because they have been displaced by superior equivalents, and others because they are found to serve no real need. Nevertheless, it is already possible to see some distinct and promising families emerging, and even if the current examples disappear they are likely to replaced by better ones.

There is, of course, considerable overlap as well as distinct differences between personal and corporate knowledge management. Both start from a heightened awareness and better understanding of knowledge, and a more considered and systematic approach to intuitive, everyday processes. Many corporate knowledge activities and tools can be used for personal knowledge purposes, and doing so usually brings corporate benefits as well. Foresight and hindsight workshops concentrate on combining knowledge and information from a group of people to make a whole that none of the participants could have produced by themselves, but they are also opportunities for individual reflection and learning, and individual insights can be valuable contributions to the group process. A corporate knowledge base can be an ideal place to store information of personal interest – well organised, easily accessible, and secure – even if there is no immediate expectation that it will be useful to anyone else. The success of mentoring and communities of practice depends just as much on participants wanting to learn and develop their own careers as on their wanting to pass on knowledge and benefit the organisation: the personal and corporate interest are two sides of the same coin. Most corporate knowledge management systems rely heavily on grass-roots effort, and personal benefits are a vital source of the motivation needed to make them succeed.

Developing personal expertise

There is no fundamental difference between the techniques that are helpful in learning from personal experience and those used in group learning processes such as hindsight reviews: systematic review of events, reflection, uncompromising search for true causes, incorporation into larger structures of understanding, and recording of lessons learned. However, personal learning can also use techniques that are rarely practicable in a group context.

As we saw in an earlier chapter, learning can be thought of as the transformation of information into usable knowledge by recognising patterns and relating it to pre-existing understanding – in other words, extending, enriching and strengthening the mental models of the world that lie behind everything we do. In the words of a survey of learning research published by the American National Academy of Science:

To develop competence in an area of inquiry, students must: (a) have a deep foundation of factual knowledge, (b) understand facts and ideas in the context of a conceptual framework, and (c) organise knowledge in ways that facilitate retrieval and application.

As we saw in an earlier chapter, experts – including all qualified professionals, to varying degrees – differ from novices in several ways. They:

- Notice meaningful patterns in information that novices miss
- Possess a great deal of content knowledge, organised in ways that reflect a deep understanding of the subject
- Relate knowledge to the circumstances in which it is applicable rather than remembering it as isolated facts or propositions
- Can retrieve knowledge flexibly and with little conscious effort – it comes to mind automatically.

The fundamental particles of mental capability are much the same for experts as for novices – they both have a short-term memory capacity for about seven items, for example – but expertise allows these to be used much more efficiently, enabling much higher performance. In one experiment, for example, electronics technicians reproduced large parts of a complex circuit diagram after only a few seconds' viewing, while novices could remember only a few components. The difference is that the experts were able to recognise – and only needed to remember – that the circuit was an amplifier with certain characteristics, and when recalling it they were able to fill in the details from their prior knowledge of amplifier circuits, whereas the novices had to remember individual components. In another experiment, expert and student physicists were asked to describe what approach they would use to solve a physics problem. The experts typically spoke about physical principles and laws, why these were applicable, and how they could be used, whereas the students simply described the equations they would use. This is fine, provided they have been taught the relevant equations and have correctly identified which ones to use, but is a fundamentally less flexible and robust approach. It is similar to the familiar difference between the novice cook's need for step-by-step recipes and the expert chef's recipe-less and apparently cavalier approach. In a third experiment, expert and novice teachers gave quite different accounts of what they had seen in a videotape of a classroom lesson. The experts described how the students were setting about their work, made inferences about their levels of ability, knowledge and interest, and supported their conclusions by reference to specific observations; the novices could only make remarks such as 'They're getting ready for class, but I can't tell what they're doing' and 'It's a lot to watch'. It is not difficult to see parallels for all of these situations in professional practice.

Research like this has removed much of the mystique surrounding expertise, giving insights into how people can learn – and be coached – to become experts more quickly. More importantly for our present purposes, it shows how we can become more expert ourselves. It shows that we need to accumulate information and practical experience from our work, from talking to other people, and from reading; to think about what we have experienced and learned in leisure moments, as well as in the rush of the current job; and to apply our knowledge to increasingly challenging situations.

Teaching and writing can help learning, too. Whether informally as a mentor or formally as a guest lecturer or author of an article for a professional journal, the need to organise our thoughts (perhaps helped by techniques such as mind mapping) and examine them critically develops our own understanding. As the old saying has it, 'How do I know what I think until I hear what I say?'

Building a bionic memory

In the knowledge manager's ideal world all codified knowledge is kept in the corporate knowledge base, accessible to everyone and used by everyone. But in the real world we all keep private stores of material we find useful, but do not want (or cannot be bothered) to put into 'the system', where it would be harder to find, might get deleted, or might be misused.

There are several software tools that can help with this. EverNote and Microsoft OneNote are both designed principally to store, organise and retrieve clips from documents and websites. They can also hold notes that the user types in, and even – with a graphics tablet – sketches and handwriting. They have a variety of tools to make this as easy as possible, including customisable folders and tags, full-text search facilities, drag-and-drop insertion. They even append an automatic link to the documents or websites where material came from. It remains to be seen how successful they will be; relatively slow take-up suggests that they need to improve to become ubiquitous.

Personal wikis are potentially more flexible and easier to use in some respects, especially when the majority of content is written by the user rather than copied from other sources. Costs are minimal: a corporate wiki installation can easily be configured to support multiple, private wikis (a so-called 'wiki farm') allowing personal wikis to be created freely.

Weblogs, usually shortened to 'blogs', are an alternative and increasingly popular form of web-based notebook. As their name implies, they are basically diaries or journals with a chronological structure (blog software normally provides a calendar for navigation), but they are often used as knowledge repositories as well. Their corporate uses are limited, and in a practice context personal wikis can be a better basis for personal knowledge stores, too. The topic-based structure of a wiki makes information retrieval easier than the date-based structure of a blog, and most of the things blog software can do, wiki software can do too.

There are useful steps that people can take to make the personal information stores on their own computers more accessible, even without resorting to new software. Most people store links to websites they like in their browsers ('Favourites' in Internet Explorer, 'Bookmarks' in Firefox); fewer realise that for years Microsoft Windows has included the facility to create 'shortcuts' to local files, making them accessible from two or more locations without the need for multiple copies. Stored in folders labelled by topic, shortcuts enable people to create an elementary knowledge base of their own from files scattered anywhere in their own or the corporate system. Windows has also included a workstation indexing service, but until recently few people have found it worth using. This has been greatly improved in the new Windows Desktop Search, which makes it possible to search the documents stored on a personal workstation instantly from the taskbar.

Facilities like these promise to improve progressively with each generation of operating system, and other technological advances are further expanding the possibilities for PKM. Pocketable flash memory devices in various forms holding several gigabytes, and micro hard drives holding much more, are already commonplace. By next year I expect to be able to carry my entire personal e-library of over 25 000 documents in a flash drive, and refer to it anywhere I have access to a computer. Now that ways have been found to run software from flash drives I may even be able to take a full-text index with me, too, so that I can find anything I want as quickly as I can at my own desk. Futurologists are already talking about the possibility of wearing miniature cameras and sound recorders and recording everything we ever see or hear. The scope for building bionic memories looks set to grow and grow.

Chapter Fourteen
Synergies

One of the commonest mistakes in knowledge management is to think about it as a set of separate tools and processes rather than as an integrated system with business objectives. Another is to concentrate on IT tools and ignore human activities such as hindsight review and mentoring. These are serious errors: synergy between tools and activities can be a great value multiplier. Various opportunities for this have been mentioned in previous chapters, but it has so much to offer that it merits recapitulation. In this chapter we will briefly revisit some of the most fruitful relationships: between the various IT tools, between the processes of creating and sharing knowledge, and between communities of practice, knowledge bases and mentoring.

IT-enabled synergies: networking directories, knowledge bases and business systems

IT-based information resources have a patchy record. Many are well liked, in constant use and self-evidently real business assets, whereas many others are disliked, little used, and largely valueless. The huge variety of systems – from accounting systems and payroll databases to wikis and networking tools – and of implementations and circumstances of use obscures the causes of success or failure (which is perhaps why the failures continue). However, some indications can be teased out from knowledge audit results and anecdotal evidence. There are several factors that at one end of their scale predispose towards success and at the other towards failure; some of the more important of these are summarised in Table 14.1.

These can interact to both good and bad effect. Extensive and useful content encourages frequent use, for example, and this brings familiarity and reduces the adverse effect of poor navigation. Thin content, on the other hand, reduces frequency of use and magnifies

	Success	Failure
Use	Necessary	Discretionary
	Frequent	Infrequent
Content	Extensive	Sparse
	Meets real user needs	Meets only imagined or management needs
Navigation	Intuitive	Needs to be learned
	Highly visible	Needs to be looked for
	No scrolling needed	Scrolling needed
	Single mouse-click	Multiple mouse-clicks
	Direct	Indirect
Change	Frequent	Infrequent

Table 14.1 Factors predisposing towards success and failure.

its adverse effect. When users are able to contribute material directly themselves, as they are in a wiki, more content attracts more users and more contributions in a virtuous circle, and vice versa. In the best cases success snowballs, while the worst enter a spiral of decline that can be difficult to reverse. The value of content and the ease with which people can find what they are looking for are particularly important, and synergies can be exploited to help with both.

Knowledge systems start with two unavoidable handicaps in comparison with accounting and other business-critical systems: their use is usually discretionary, and it is difficult to justify investment in extensive content until they have proved their worth. This is a classic chicken-and-egg situation, and managers and system designers need to work hard to break the impasse. They can do this by exploiting content from existing business systems and designing the new systems to be as intuitive and user-friendly as possible.

It is rarely necessary to develop all the content of networking tools and knowledge bases from scratch. Most organisations have a considerable amount of relevant material scattered around in existing business systems and elsewhere. Imports can provide an invaluable nucleus, with the incidental benefit that incorporating material into the new systems can make it more visible, accessible, and hence valuable. Much of the content for networking tools, for example, including names, photos, contact details, qualifications and projects worked on, can usually be drawn from HR and MIS databases; project directories can usually draw basic information from existing databases in the same way. That is enough to make the new systems immediately useful, and leaves staff free to focus on adding more interesting and higher-value information about experience, personal interests, the design philosophy behind projects and so on.

Ideally connections between existing and new information systems should be made using live links, so that when business databases are updated the changes are reflected immediately in the knowledge systems (a technique, incidentally, used increasingly on the Web, where sites that aggregate material from multiple sources are proliferating). This enables them to benefit from the high maintenance standards required in business-critical databases, while the added visibility they give to the data makes it more likely that any errors will be spotted quickly. If the new tools provide a more user-friendly and efficient way of accessing data that is already in frequent use, so much the better: it will attract users to the new systems, and they will soon start using them more fully.

Exploiting existing material in knowledge bases calls for a different approach. This is more often scattered between separate text, pdf and other document files than conveniently collated in databases, and human intervention is unavoidable. It is rarely satisfactory simply to provide links to the original documents; the extra effort required to open separate files, minimal as it is, is almost a guarantee that they will rarely be read. Existing resources are best regarded as quarries from which text and illustrations can be extracted and reshaped to suit their new context, and as supplementary material to support summaries with chapter and verse. The process of selection, content extraction and adaptation is labour-intensive, but less so than writing *ab initio* – and it helps to ensure that valuable material is not ignored, and to avoid needless reinvention of wheels.

The user-friendliness of IT-based knowledge systems depends in part on the visual details of interfaces – page layout, fonts, colours, use of icons and so on – but much more on how they relate to each other, and to users' mental models and expectations. One aspect of this is the way that the content of the individual systems is organised, as we saw in a previous chapter. Another is how they are interlinked. Imaginative linking can add greatly both to their user-friendliness and to their value by:

- *Signposting* the existence of information that might otherwise be forgotten
- *Encouraging* users to use more of the available resources
- *Making relationships visible* that might otherwise go unnoticed
- *Providing multiple routes* to information
- *Making navigation more intuitive* and less dependent on familiarity with system-specific features that need to be learned
- *Reducing search failures.*

Wikipedia provides an excellent example of this kind of linking in action (and, incidentally, the ease with which it can be achieved in

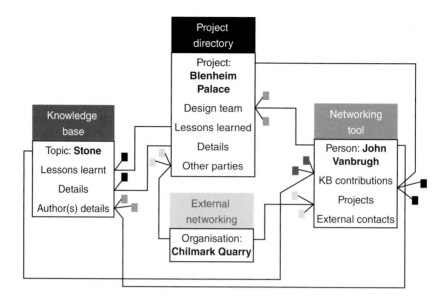

Figure 14.1 Synergies between software knowledge tools – internal and external networking directories, a knowledge base, and a project directory.

wikis is one of the reasons why they make a good framework for knowledge systems). The overall effect is to give users better information more easily and quickly, and make using documented knowledge a more rewarding experience. A designer starting work on a new shopping centre, for example, might go into the knowledge base to look for general guidance on the sector (native knowledge base content), see a list of recent retail projects, click through to one of them (project directory content), look at who worked on it, click through to the project leader's personal page (networking tool content), send an email, then click back to the project, see that a particular material was used, click through to technical information about it (knowledge base again), find other projects where that was used, look at those, and so on, being reminded on the way about (and given immediate access to) information on markets, clients, materials, suppliers, lessons learned, knowledgeable colleagues and other relevant material. Figure 14.1 illustrates some of the ways in which links between knowledge systems can make them mutually enhancing, and add value to them all.

Creating and sharing knowledge: foresight, hindsight and knowledge bases

The principal purpose of foresight is to ensure that the best expertise and most imaginative ideas are brought to bear on projects before decisions foreclose options, and the principal purpose of hindsight is to learn lessons from experience and reduce the number of mistakes that are repeated. Both are creative enterprises that, at their best, produce knowledge that is new, potentially valuable and unique to the company – precisely the kind of content that is most worth sharing across the whole practice. This makes foresight and hindsight reviews among the most valuable sources of content for a knowledge base.

Recording imaginative design ideas and lessons learned in a knowledge base solves a problem for the reviews, too: how to document the outcomes. A hindsight review usually gives rise to several lessons learned, on unconnected topics, and there are obvious disadvantages both in recording them in separate documents and in combining them in a single report. With either approach, the accumulated records become increasingly hard to use as the number of reviews builds up, and would-be users have to collate and reconcile information from more and more sources. It is much better to append each lesson learned to others on the same topic in the appropriate page of a knowledge base. There, it is immediately obvious when a lesson repeats or appears to conflicts with a previous one, and it can be recorded appropriately – either limiting it to an endorsement, or explaining the discrepancy. The result is a page where knowledge base users can confidently expect to find all the available information on the topic, in a form that makes it immediately usable.

Multiple synergies: communities of practice, knowledge bases and mentoring

Communities of practice can have constructive relationships with most other knowledge tools and activities. They are well equipped, for example, to:

- Create authoritative content for knowledge bases – which provide them with a publishing platform
- Take responsibility for moderating knowledge bases – a process which helps them to keep in touch with lessons learned
- Act in a quasi-mentoring role for their less expert members – who can help ensure that their knowledge base contributions are pitched at a suitable level for the users who most need them
- Help capture the wisdom of departing seniors – which in turn will help to develop them as centres of expertise.

Practices that are too small for communities of practice to be viable (and larger ones that have none) often have recognised experts or specialist knowledge facilitators, and there can be similar opportunities for synergy between their work and knowledge systems and activities. Services such as WSP's technical 'help desk' and Bovis Lend Lease's *iKonnect*, for example, can benefit considerably from networking tools (which can help them find answers to questions), and answers to the queries they receive can make good material for a knowledge base.

Value-adding synergies such as these become apparent only when managers take a systemic view of knowledge management. It really does pay.

Part Three
Knowledge Management in Practice

Chapter Fifteen
Introduction to
the Case Studies

The case studies

The success of knowledge management turns on how well the choice
and execution of strategies, tools and processes fit the organisational
context. The devil really is in the detail, and nothing makes this so
clear as real examples.

The case studies that follow show how over a dozen design practices and other organisations in the construction sector have set about
managing their knowledge. They vary in size from an architectural
practice with under 40 staff in a single office to an international engineering consultancy with over 7000 spread across 70, and large plc
clients such as oil company BP and airport operator BAA. They are
variously design led and process led, managerially centralised and
managerially dispersed, with IT skills varying from amateur to professional. Some of them were thinking systematically about knowledge
and knowledge management for the first time, some were refreshing
long-established practice, and some were simply experimenting with
specific tools or processes. The cases describe in detail how they
translated basic principles and techniques into concrete reality, and
discuss the factors that shaped their thinking, the problems they
encountered and, in many cases, how well they succeeded.

The first nine cases follow the experience of design practices that
participated in a collaborative research project called 'Spreading the
Word' between 2003 and 2005, which I set up and led. Some also
follow subsequent developments to the time of writing in 2007. The
other five cases are taken from one of my earlier projects, which
focused specifically on learning from experience and the use of
foresight and hindsight. They are all rich sources of inspiration and
practical ideas.

Recurring patterns

Several patterns recur in the nine cases that look at knowledge management as a whole. These will all be familiar from earlier chapters, and they echo experience in other industries; they are not peculiar to these particular practices.

Leadership

Leaders need to lead: partners/directors need to be both fully committed to and visibly engaged in knowledge management to get the best from it. They are the only people who have a sufficiently clear and realistic view of business objectives to align knowledge systems fully with them, and who can ensure that appropriate resources are allocated to knowledge initiatives. They are also the only people with a wide enough span of authority to make knowledge a pervasive consideration in management – to ensure, for example, that staff appraisals and time budgets support knowledge initiatives, and that effort is not diverted to apparently urgent but fundamentally less important work. And their involvement must be visible, because people believe the lead that top management gives through its actions, not what it says.

Costs and benefits

Business leaders are increasingly recognising that knowledge management as such does not need cost–benefit analysis: it is becoming a precondition for staying competitive. It is in any case impossible to calculate either overall costs or benefits – both are simply too diffuse and uncertain. If a plausibility argument is needed, knowledge audits show consistently that staff in design practices believe they spend a third of their time reinventing wheels, doing rework, or searching for information. Reducing this even fractionally by improving learning and knowledge-sharing will pay for all the costs of knowledge management, effectively delivering the other business benefits for free. Evidence from other industries suggests that this is typical.

The main difficulty that practices encounter when they embark on an initiative to improve their knowledge management is not in taking a decision of principle, or even in allocating a budget for consultancy support or software, but in sustaining the attention and active engagement of managers, at all levels, who are overwhelmed by short-term urgencies. Without this, progress is at best severely constrained, and may be impossible.

It *is*, though, important to consider the business value of specific activities. If this is not done, it is easy to lose sight of their purpose, and steps need to be taken to refocus them.

Scale

Different sizes of organisation need different mixes of tools and processes. Knowledge bases of one kind or another – even if they serve only to make basic information about people and projects readily accessible, and act as a communal notepad – are valuable, regardless of size. We forget things, we cannot ask colleagues when they are out of the office or they have left the practice (and they can forget, too), and we all waste time searching for material that is not where we thought it was. On the other hand, networking tools are largely superfluous for a few people working in a single office, and communities of practice only start to become viable when staff numbers rise above 100 or so.

The larger the practice, the larger the potential benefits of pooling knowledge. Without effective knowledge management, a large organisation is little more than a collection of small ones, and it loses much of the competitive advantage its size could bring. On the other hand, small practices can often develop new strategies, tools and processes for managing knowledge more quickly than large ones, and this can help them keep ahead and continue to compete successfully.

IT

People are always more important than IT, but it is an indispensable enabler. The most valuable knowledge is usually tacit, and can be shared only directly, person to person; routine and trivial knowledge is often more effectively shared person to person, too. Much of this can be achieved without IT – for example by mentoring, in foresight and hindsight reviews, and by good design of the workplace. However, there is no alternative to IT tools for connecting people who do not work close to each other. They are indispensible for facilitating practice-wide networking, as virtual homes and publishing platforms for communities of practice, and for storing codified knowledge resources in an accessible way.

Expertise

Knowledge management is not easy, and common sense is not enough to make a success of it. Expertise and experience are as indispensable in this field as in any other aspect of professional practice and management. These can be acquired only by study, by trial and error, or by working with an expert. The opportunity costs of senior staff time and delay in making knowledge management effective are so high that expert help is usually a good investment.

Preparation

Don't rush in. Knowledge management is a complex area of expertise, and it can take months of thought and discussion for managers to understand what it implies for an organisation and their own priorities and decisions, and to equip them to plan and lead knowledge activities successfully.

Patience

Don't expect to see benefits too soon. It often takes longer than people expect to implement new processes and IT tools, and that is only the start. It can take two years or more for a majority of staff to learn how to use them, see their value, and develop new working habits, and longer still for benefits to become clear. Some initiatives may disappoint, and need to be reviewed, redesigned and restarted. Until knowledge systems have proved themselves and become part of the culture, they need to be given continuing support from visible leadership, publicity, active evangelism, and complementary policies in other aspects of management. Only the simplest of the initiatives started during the Spreading the Word project could be said to have been completed by the time it finished, and in all the practices that took part knowledge management remains an evolving story today.

Chapter Sixteen
Case Study: Aedas

In 1993 a round of impending partner retirements at architects Abbey Hanson Rowe prompted a strategic review of the business. This crystallised a number of issues, including:

- The difficulty of managing a partnership in which all the owners also wanted to be managers
- A lack of professional management expertise
- Partners having too little time to practise architecture
- The fact that, with 125 staff spread across four offices, the practice was neither truly regional nor truly national.

The outcome was two rounds of mergers involving a total of four practices. In 2003 these finally became Aedas, an international practice with the scale necessary to support a professional management team. Its approach to managing information and knowledge has been evolving ever since, and further expansion has made Aedas the fourth largest architectural practice in the world.

This case study follows its progress through to 2007.

Practice profile (2007)

Staff: 600 UK (1900 worldwide)
Offices: 9 UK (26 worldwide)
Services: Architecture, interior design, landscape and environmental design, building and land surveying, imaging, and workplace, access and CDM consultancy
Web: www.aedas.com

Case study themes: workplace design, knowledge audit and software systems

Starting points

It was clear from the start of the mergers in 1993 that greater size would bring business benefits only if the whole could be made more than the sum of the parts. Today, that would be seen as a classic case for knowledge management, but at the time that was an almost unknown concept. Abbey Hanson Rowe had just two librarians, its only computerised information systems were payroll, job cost reporting and a homespun marketing information system known as the MKIS, and the practice's only significant knowledge asset was the experience of a number of long-serving staff.

The first merger made it imperative to develop better systems to help integrate the businesses, and to support more complex and rigorous management. The practice began to develop a new management information system based on Microsoft Access, and this matured progressively over several years to replace the job cost reporting software and the MKIS. The MIS and its underlying IT infrastructure have continued to develop, and today (no longer based on Access) it comprises a range of business, quality and information management tools, and runs on an intranet that links all nine UK offices.

By 2004 Aedas had reached a stage in its development where further enlargement was no longer a priority (though it has since expanded further, mainly overseas), and the focus turned towards design quality. This created a new imperative to share design knowledge as effectively as the established systems shared management information. Aedas's initial response was to create the 'Aedas Studio', a new workspace in the London office designed to be the crucible for creativity and design skills throughout the practice. This was followed in 2006 by a major new knowledge management initiative, still under way at the time of writing in 2007, which is introducing hindsight reviews and an integrated set of IT tools designed to facilitate learning and knowledge-sharing. The new tools will also help support personnel management and marketing, aspects of management that are not well served by the MIS.

MIS

From its beginnings as a tool designed solely to support business management, Aedas's MIS gradually evolved into a multifunction intranet with a variety of information-sharing roles, intended to serve the needs of all the staff. It has recently been rewritten (using a team of three full-time in-house IT staff) in order to bring the IT infrastructure up to date and provide a better basis for future developments while continuing to support the practice's key management databases and reporting functions. It now includes:

- An address book and contact database
- A staff database

- A project database
- Time sheets, holidays and expenses
- Job cost reports
- Resource charts
- A CV database
- An image database
- A press cuttings database.

Other features include:

- Newsreel, a simple scrolling newsreel that allows news items to be rolled out as they arise
- Project Showcase, which dynamically combines paragraphs of text from the project database and images from the image database to show what is going on throughout the practice
- 'Active content pages' for each of the main market sectors, which have some of the basic functionality of a wiki and open the way for creating a rich technical and design knowledge base
- 'Dashboard' components that provide a quick visual indicator of the health of the business in a number of different ways. They can show, for example, staff without work, staff with future leave booked, and future submissions due. Additional dashboard components have since been added that users can configure to suit their individual needs.
- Links to other SQL databases such as the Sun Accounts software and the Conisio document management system
- Access to external information resources such as the Barbour Technical Library and to a range of system services including website usage statistics and systems monitoring utilities
- A comprehensive help system
- A search function. This indexes the metadata in files with information content – notably Microsoft Word, Excel and PowerPoint files, and pdfs – as well as the full text of documents, so that searches include attached files as well as text on the intranet pages themselves.
- Menus that are dynamically constructed to suit individual users' access rights.

The active content pages (ACP) allow web pages to be created with no specialist training or HTML skills, allowing architectural staff to create new pages, insert links to other pages, attach files, include keywords to facilitate searching, and reposition existing pages within the intranet structure.

Figures 16.1 to 16.3 show the evolution of the MIS from 1996 to the present day.

Evolution of a management information system

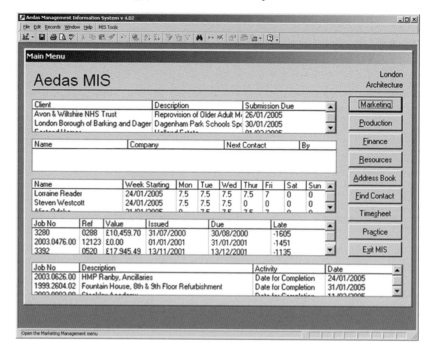

Figure 16.1 The MIS in 1996: a Microsoft Access database, with information on marketing, production, finance, resources, addresses and time sheets.

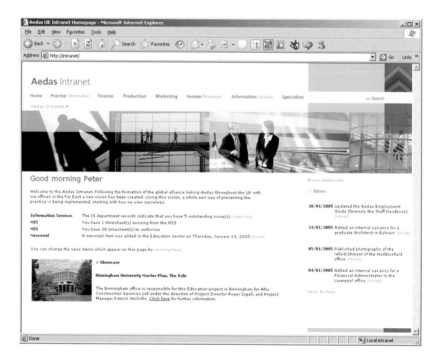

Figure 16.2 The MIS in 2002: a new intranet-based system, with news and personalised action prompts, and links to practice information, finance, production, marketing, HR, information services and specialists.

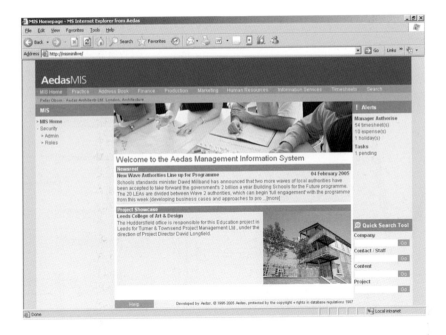

Figure 16.3 Today's intranet-based MIS, comprehensively rewritten using .Net and SQL technologies, and with an updated interface.

Aedas Studio

It is easy to decide to improve design quality, but much harder to decide how. How do you:

- Improve design quality in a company of over 600 staff across nine offices (let alone in a worldwide group three times that size)?
- Maintain the client base and the business while changing the philosophy of the company?
- Create a high-profile architectural practice that can sit alongside the best, while continuing to maintain a high level of service to existing clients?

As a first step, Aedas decided to headhunt an experienced design director, and he joined the practice in late 2004. With his help, they developed two options: to spread the design director more or less equally across all nine offices, or to create an 'ideal' design studio in one office to act as a nucleus for other offices to use as appropriate. The second quickly emerged as the winner. Aedas judged that input from a design director would be too thin and insufficiently continuous if spread across nine offices, and existing clients might not all be supportive. With a single centre of excellence, on the other hand, it would

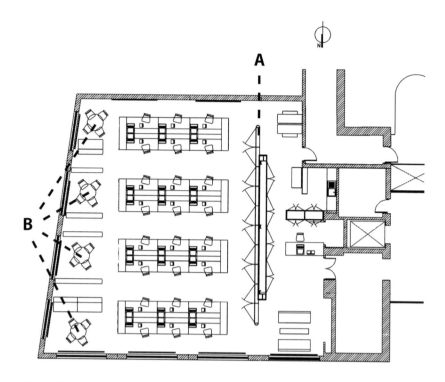

Figure 16.4 The plan of the Aedas Studio. (A) is a magnetic pin-up wall; (B) are break-out spaces.

be possible to develop new ways of working in an environment designed for the purpose and with a strong design focus, working selectively with supportive clients. Successful procedures could then be rolled out across the practice, and the studio facilities could be made available to other offices to use as much (or as little) as they wished.

The London office was enjoying a period of organic growth at the time, and had recently taken on additional space, so it was an easy decision to fit out the new space as a bespoke studio (Figure 16.4). The main features of this are:

- A prominent free-standing magnetic pin-up wall (A) for displaying work within the office. (The wall also doubles as a storage unit.)
- A series of breakout spaces (B) where informal meetings can be held. These are divided by metal filing cabinets whose backs provide pin-up space for the breakout spaces.
- Directors moved out of cellular offices into the main open-plan space to maximise accessibility and involvement.

- Cordless phones for the directors so that they can go somewhere more private for calls if necessary.
- A wireless network so that senior staff can move around the office with their laptops while remaining connected to the network.

The layout of the new Studio (see Figure 16.4) creates a non-hierarchical space in which:

- Work is always prominent, and changing.
- All staff have an opportunity to contribute to design.
- Everyone is aware of what others are doing, both through seeing the work and through hearing what is going on.
- There is a sense of involvement, encouraging the exchange of ideas and the sharing of knowledge.

The Studio is equipped with a comprehensive range of tools to support greater use of 3D modelling throughout the design process, including a model-making area with hot-wire foam cutters, substantially increased computer-modelling facilities ranging from SketchUp to Bentley MicroStation (in an otherwise AutoCAD-based practice), and improved environmental modelling software.

However, layout and equipment on their own achieve nothing: Aedas recognised that the success of the Studio depended on the creativity of the staff and the rigour of the analysis and review being applied in the design process. They have therefore started to capture this process in 'storyboards', which can be used both as training aids and to communicate with clients and with the other offices. In the future they also plan to use foresight and hindsight reviews to trap the results of design reviews so that lessons can be learned and fed back into the design process throughout the practice.

The Studio produced a string of competition-winning designs in its first year and, not surprisingly, Aedas considers it to be an outstanding success.

Knowledge audit

Aedas decided in early 2006 to take stock of its progress in managing knowledge with a formal audit of the UK business commissioned from an independent consultant.[1] Carried out over six weeks, this drew evidence from:

---

[1] The author.

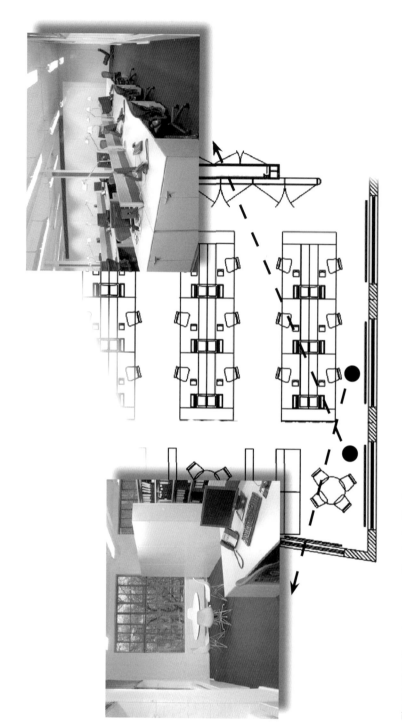

Figure 16.5 Two views across the Aedas Studio.

- Material from Aedas's involvement in the Spreading the Word project, and subsequent internal documents relevant to knowledge management
- Visits to four offices
- One-to-one interviews with 16 staff (including the managing director and eight other directors) and group discussions with 16 others
- A questionnaire circulated to all staff and completed by over 26%.

The interviews gave a rich insight into the concerns and perceptions of staff at all levels, and showed what issues would be worth investigating further. The questionnaire picked up on these, providing quantitative evidence to back up or modify the impressions gained from the interviews. It included questions on:

- Respondents' location, length of service and other personal characteristics, to enable differential analysis of their responses to other questions
- Learning from experience, including questions about issues such as the value of project files as sources of information, personal learning habits, and participation in team reviews at the end of projects
- Talking to colleagues, with questions exploring the range of people's networks and the opportunities for overlooking, overhearing and casual interaction in the office
- The active content pages on the MIS, at the time the only practice-wide codified knowledge resource
- Membership of CoP-like groups (Aedas had no communities of practice as such at the time)
- Mentoring
- Aedas Studio.

The audit showed that, with few exceptions, office directors and other senior managers were all highly conscious of the importance of knowledge and the potential benefits of better learning and sharing. Some had taken local initiatives to try to improve them, but with limited success. Pressure of work was a major obstacle. For example, although 95% of directors (and 80% of other staff) thought that the time and effort involved in systematic post-project reviews was repaid in future time savings, in better design, or in other ways, very few were done in practice.

The questionnaire returns showed that face-to-face knowledge-sharing was confined largely to colleagues people had worked with previously. Most staff had a fair idea of what was going on and 'who knew what' in their own office, but only directors had more than a few, occasional contacts elsewhere. Only a minority found the staff database in the MIS useful for locating people with specific expertise and experience. Many people said they looked back at files on similar projects before starting new ones, but they evidently did so more in hope than in expectation; most found them never or only occasionally helpful. Two-thirds claimed to reflect regularly on their projects post-completion in order to learn lessons from them, but few ever discussed projects with colleagues after completion, and lessons learned were rarely recorded in a way that made them accessible to others. Not surprisingly, staff thought they spent an average of 18% of their time reinventing wheels and doing rework.

The sector active content pages were a disappointment. Content was sparse, chiefly because write access was limited to the chairmen of the sector focus groups, and they rarely had time to create any. As a result, working architects found the system unrewarding, and it was little used. Less than 20% of staff below director level even knew it existed.

The story on the practice's largest knowledge initiative, the Studio, was much more encouraging. Only a minority of staff had had the opportunity to work with the Studio on a project, but they were over-whelmingly positive about the experience: nearly 90% found the experience 'very' or 'somewhat' stimulating, and 60% thought the collaboration had resulted in a better building. The facilities in the Studio for displaying work in progress and for breakout meetings, and the collaborative and non-hierarchical working practices, were widely admired. However, the record was tempered by some resentment on the part of staff who had not benefited personally from contact with the Studio. It was quite widely perceived as taking an unfair share of the most interesting projects, and having the luxury of less demanding business targets. It is immaterial whether perceptions like these are justified or not; they have the same effect in either case.

Overall, the audit concluded that 'learning and knowledge-sharing in Aedas are largely personal, spontaneous and unsystematic', and the practice evidently had a long way to go to become much more than the sum of its parts in knowledge terms.

Emerging knowledge systems

The audit findings convinced the Aedas board that they should launch a major initiative to improve the practice's management of knowledge. With the help of an external consultant,[2] a strategy was developed to address the main shortcomings. This includes:

---

[2] Also the author.

- A new networking tool, to connect people who want to know something with colleagues likely to have the answer
- A systematic programme of hindsight reviews, to extract more lessons from experience, record them in shareable form, and help make design decisions more evidence-based
- A wiki knowledge base where all kinds of knowledge – lessons learned from hindsight reviews and everyday practice, articles by Aedas experts, web links and so on – can be recorded and made easily accessible.

A new design-oriented project directory has since been added to the list. The project details available through the MIS are geared towards business management, and contain little of the descriptive information that is needed in preparing bids or material of interest to designers, and no lessons learned. The new project directory is intended to complement the MIS by filling these gaps, and also to give easy access to contextual information related to the lessons learned that are recorded in the knowledge base.

The intention is to let these bed down before starting on other developments such as communities of practice and a more systematic mentoring programme.

When the current version of the MIS was designed, the expectation was that the active content pages would allow it to evolve into a knowledge management tool as well as a business management tool. The audit showed that this would be a mistake. The MIS serves the needs of management – and particularly senior management – very well, but it is generally disliked by grass-roots staff. They find it user-hostile, as well as unrewarding as a source of information. To avoid the stigma the MIS had acquired, it was clear that the new knowledge systems would have to have an entirely different look and feel.

The implementation plan therefore proposed that the networking directory, knowledge base and project directory should all be based on a wiki software platform. It was expected that this would minimise costs and make it easier to differentiate the new systems from the MIS, while providing all the desirable functionality: write access from all desks, powerful and intuitive navigation and search facilities, easy quality control, and the flexibility needed to adapt to changing needs. The use of a common platform would facilitate rich hyperlinking between the three systems, adding value to all three in the way discussed in Chapter 14, and creating a seamless knowledge space. In the event the IT team judged that it would be more expensive to familiarise themselves with new software and develop interfaces to connect with data in the MIS than to extend the capabilities of the existing active content pages. This has now been done, creating a visually distinct software platform with all the important functional attributes of a wiki, while behind the scenes enabling the knowledge systems and the MIS to have common data structures and share content transparently and reliably.

At the time of writing, all three of the new IT systems have been through several iterations of design, testing and improvement, and they are about to be launched.

The hindsight programme, though, has been held back by pressure of work and a shortage of senior staff time, and it has yet to get off the ground. It will be the next focus for attention.

Commentary

Aedas's experience shows that even with clear support at board level and an engaged and enthusiastic managing director it is hard to stop the urgency of completing today's project and winning tomorrow's trumping the strategic importance of knowledge activities for attention. This is particularly so in organisations whose directors are also working professionals and shareholders: both their intellectual and (short-term) financial interests favour the projects. It took Aedas nearly six months to progress from a board decision to the start of active work on its new knowledge strategy, and another year to get the new software tools ready for launch. That is not untypical for a large practice.

The knowledge audit was an important step. It provided the hard evidence needed to remind the board of the benefits of better learning and knowledge-sharing and to secure agreement to more investment, and it provided a road map for moving forward. The distractions of expansion and a booming construction market made its psychological effect disappointingly brief, but it left a lasting legacy in the framework it provided for developing a knowledge strategy and an implementation plan matched to the practice's needs and circumstances.

With hindsight, subsequent events were, perhaps, predictable. Aedas has an excellent record of creating good software tools to support its work – first the MIS, and now the new networking directory, knowledge base and project directory – and that part of the plan has gone ahead successfully, if slowly. However, it has yet to find a way to divert enough of the office directors' and middle managers' time away from projects to create a critical mass of content, or to induct staff in the use of the tools or in processes such as hindsight review. This is a challenge that many other organisations face.

An MIS is not strictly a knowledge system, but the parallels with tools such as knowledge bases are close enough for Aedas's experience with theirs to be instructive. The obvious practical value to senior managers of having up-to-date business performance data at their fingertips created a favourable climate for its development, but it seems unlikely that the MIS would have become so powerful so quickly without the enthusiasm and drive of one man (now the managing director). When the bulk of management attention is elsewhere, individual champions make a crucial difference.

It is unlikely to be an accident that Aedas Studio has been Aedas's most successful knowledge initiative so far, albeit one that appears to have faltered with the departure of the founding design director. When there is a will to make things happen, but a difficulty in *sustaining* attention, one-off, buy-and-forget investments are the most likely to succeed – provided, of course, that they are capable of delivering what they promise.[3] Aedas Studio might well not have succeeded if it had not been essentially self-sustaining, with the effort and enthusiasm of its own staff to drive it and a financial mechanism (based on cross-charging for its contribution to projects) to support it.

---

[3] The allure of buy-and-forget is one reason why there is still a market for expensive software that promises to provide a complete knowledge management 'solution'.

Chapter Seventeen
Case Study: Arup

Arup has been conscious of the importance of knowledge throughout its 60-year history. Founder Ove Arup spoke in the 1940s about how difficult it was even then for engineers to 'become familiar with the complete range of modern technical possibilities', and of the need for design practices to develop a 'composite mind', sharing knowledge across the organisation. The practice has had an active programme of what we would now call knowledge management for many years, and it has more experience in the field than any of the other practices involved in the Spreading the Word project. Arup used the opportunity to improve the business focus of its skills networks – its name for communities of practice – and to explore the use of storytelling and workshops to create, share and codify knowledge. This is Group Knowledge Manager Tony Sheehan's description of their experience.

Practice profile (2005)

Staff: 3000 UK staff (7000 worldwide)
Office: 19 UK offices (70 worldwide)
Services: Engineering design, planning and project management
            services in all areas of the built environment
Web: www.arup.com

Case study themes: linking CoP activity to business, storytelling, and knowledge-sharing workshops

---

Starting points

Our KM tradition can be traced back to the founder of the firm, Ove Arup, who had a very strong belief in the importance of sharing knowledge from its beginnings in the 1940s. As a result of this strong drive from the top, various good KM practices evolved within Arup at an early stage. Lessons learned, for example, have been routinely captured and shared since the 1960s, and a culture of sharing and knowledge reuse is a key feature of an increasingly global firm.

Knowledge management practice was reviewed in 2000, and we introduced various new practices including communities of practice, a revised intranet, and an approach to managing electronic knowledge targeted at reducing information overload.

We recognised that to deliver business in today's pressured work environment we required an investment in world-class knowledge management practices. Clients need to consistently access the best knowledge and receive the best of Arup. Knowledge management improves efficiency and makes time for creative thinking by building on the successful practices of the past to prepare the firm for the future. We link our KM activities to business impact to ensure that practices are appropriate, and that investment can be justified.

Given our goal to achieve innovation and creativity in our projects, our KM approach maintains an element of standardisation, procedure and IT, but focuses on these elements far less than in many organisations. Instead, we seek to combine people, process and technology to support a less structured approach, encouraging innovation and flexible working practices across many sectors.

By 2004 we had already established a number of good practices in KM, including:

- Arup People, an award-winning system to find experts within the firm
- Arup Projects, to capture lessons learned on projects and to access key project data and images
- Arup Networks, communities of practice linking people around the world who are working on separate projects but are united by a common interest
- An intranet system, using a powerful search engine to cut across organisational boundaries and access appropriate best practices
- An appraisal system to support appropriate behaviours
- Processes and procedures to consistently reapply best practice where appropriate.

Nevertheless, we were conscious that there was scope to make our KM tools and techniques work better.

Projects

As a globally dispersed firm of 7000, we recognised that communities of practice – our skills and business networks – were the key vehicle enabling us to deliver the best of the firm, and the pilot projects undertaken during the Spreading the Word project largely targeted this area. The skills networks and business sectors work closely together to ensure that the right skills and knowledge are being developed in response to business sector and ultimately client needs. Without these networks, we would offer little more than a firm 10% of our size. Building on our existing approaches, therefore, there was a desire to explore new ways of improving the impact of Arup Networks.

The skills networks are the primary means for provoking and enabling continuous cross-group technical activities to promote and sharpen our competitive advantage. Activities revolve around supporting technical skills through providing training and guidance, nurturing a culture of sharing innovative work and experiences that helps to deliver excellent projects, avoid errors and maximise the value of expertise for the benefit of Arup clients.

Within the largest of our global networks, the structural skills network (SSN), we sought to explore:

- New ways of articulating added value to businesses from network activity
- Cultivating best practice guidance through storytelling
- Workshop facilitation to initiate network activity.

Sharpening business focus

Our first pilot project assessed the impact of the structural skills network on the health care and sport businesses. The process was similar for both. Interviews were carried out with engineers working in each business sector to identify key technical issues, and to encourage an articulation of the business value created by structural skills networking activities. The interviews were followed by a series of facilitated regional workshops, which enabled key technical reports, documents and best practice to be identified.

The knowledge from these documents was disseminated through integration with the Arup project database and creation of a business-focused page within the SSN's intranet. In parallel, the whole process created active communities in each business area, and thanks to active regional network leaders the business areas are continuing to receive contributions to the present day. Integration with the business is critical to ensure valuable knowledge-sharing in the long term.

This activity resulted in numerous benefits for the health care business. It:

- Identified key people with project experience in the health care sector
- Increased efficiency through sharing best practice
- Raised awareness of 'added value', improving client focus

- Facilitated a global focus on sharing business-focused knowledge
- Ensured that technical issues with cross-business-sector relevance, such as laboratories and vibration criteria, were addressed
- Led to the development of a structural capability statement for health care.

Using stories

In parallel with developing a focus on the business, we explored:

- Storytelling as a technique for capturing knowledge, combined with
- Workshop facilitation as a technique for seeding interest groups and initiating network activity.

Storytelling is a well-established KM technique, which has been championed by the likes of Steve Denning at the World Bank and David Snowden when at IBM. At Arup, storytelling was recognised as a key approach to knowledge-sharing within communities where multiple project experiences are being discussed. The storytelling activity arose from a recognition by our project engineers that we had to create less formal vehicles for sharing knowledge. It was recognised as particularly valuable in contentious areas where experiences needed to be shared between small groups (contract disputes, for example) and in areas where agreement on best practices had not been well established. In the latter cases, storytelling serves to encourage open exploration of the issues before filtering these issues into reusable guidance.

One area where insufficient knowledge of our global project experience was recognised was visual concrete – an increasingly popular theme amongst clients. It was recognised that there was value to be gained in collating a range of our project experience in this area, both with regard to being able to draw from past technical experience and in communicating specific issues such as architect expectations of finish.

Facilitated workshops were found to be the best technique to encourage participants to share stories of their experience and explore useful methods of knowledge dissemination that would benefit the wider community. Engineers were actively encouraged to share negative experiences as well as positive ones, to raise awareness of 'what could go wrong'. We realised over time that perceived experts were not essential to these sessions – in fact, their presence almost discouraged open discussion and free-flowing stories in some cases. Their role was, however, still critical in helping to filter the outputs of workshops into reusable insights.

The workshop on visual concrete resulted in a distillation of key issues for reuse on future projects. Some of these issues were technical, some contractual; some concerned relationship and expectation management with clients. In some cases this content has been developed in conjunction with architectural clients, so that common standards can be established and better knowledge-sharing achieved.

One of the key outputs from the workshop was a commitment by the participants to provide records of project experience in a variety of formats, ranging from images of good and bad finishes to specification clauses and useful technical documents. An intranet page was created within the SSN's site to share many of the outcomes, which has since been used to improve the effectiveness of projects delivered in this area. The workshops, then, act not just as a source of content, but also as a means to galvanise a community and ensure ongoing involvement of the participants.

The process showed that storytelling can be encouraged at many levels – from informal, unstructured sessions to pre-planned presentations to seed the discussion. However, the facilitated workshop format was recognised as a valuable technique to help facilitate other network activities – whether at start-up, or as an intervention to reinvigorate declining or stagnant networks.

Future

These pilot projects have helped us to increase the adoption of our existing systems and techniques. We have sought to create a coordinated approach to knowledge-sharing between the many communication vehicles available in Arup. Face-to-face, paper and electronic methods need to be balanced to ensure that knowledge is not just captured, but developed for reuse, so we have established links between communities and general Arup newsletters in order to ensure that knowledge-sharing is maximised.

Some of the improvements recognised as a result of the project include:

- Integration of community activities with the knowledge systems to manage people, projects and best practices
- Facilitation techniques to encourage community start-up and to reinvigorate communities where activity is declining
- Improved automation of knowledge exchange from projects.

In the short term, Arup will build on this progress by further exploring how to get the best out of the people using these systems, focusing on both virtual and actual motivation and human–computer interaction.

Commentary

Arup has some of the most highly developed processes and tools for learning and knowledge-sharing in the construction industry, and there is little doubt that this is due largely to the vision and leadership of its founder, Ove Arup. He embedded knowledge awareness in the practice's DNA, and it is still reaping the benefits. As this case study shows, though, there is always scope to improve.

Even without its cultural heritage, Arup would be better placed than most practices to make a success of communities of practice. The various branches of engineering are sufficiently different to create an emotional need for social structures that enable staff to keep in touch with their professional peers, independently of the formal framework of management hierarchy, office location and project team. Arup's high standing, its involvement in an unusually large number of interest-ing projects, and its history of cutting-edge work and engagement in research all help it to attract an unusually high proportion of staff who have the interest and drive needed to make CoPs work. The practice is large enough for communities to maintain critical mass, even though only a minority of staff are active members. These are major advan-tages, which few share.

The issues that Arup addressed during Spreading the Word, though, are common to many professional services organisations, and the techniques it tried out could be used in contexts other than CoPs.

It is easy in any learning or knowledge-sharing activity to lose sight of its business purpose. CoPs can operate happily as professional talking shops and organisers of semi-social events without generating any significant business value; material is often written for knowledge bases without considering the practicalities of how it could be used; many staff induction processes are rituals that leave joiners little the wiser. In any practice, it is well worth instigating activities from time to time to refocus attention on business value, and show people how to think constructively about it and create it.

Simple instructions to 'consider business value' are doomed to failure, and presentations and articles about it in house magazines not much better. Most professionals lack the mental framework to absorb the messages or react constructively to them; they need to engage actively in a process that can help them develop one. Arup's pro-gramme of interviews, facilitated workshops and follow-up activity led by regional network leaders was well calculated to achieve this, and the most important elements of it – dedicated staff to drive the process, identification of specific issues relevant to the target audi-ence, and carefully designed events that give staff time, stimulus and opportunity to hear, think and talk about them – could be used in any practice, regardless of professional make-up or size. Only the follow-up would need to be changed to suit different organisational contexts.

Stories are one of the most natural forms of human discourse, often the easiest vehicle for making a point, and certainly one of the most memorable. It is perverse of scientists and engineers to have rejected them for so long, and they are overdue for rehabilitation. They are particularly promising as a technique for improving the dissemination of lessons learned within professional practices: the story format is especially apt for these because they derive directly from experience, and they are usually simple enough to be well within the (fairly limited) carrying capacity of a story.

After so long in the wilderness, stories need a deliberate effort to reintroduce them into professional use. People need to be convinced that it is legitimate to use them for serious purposes, and be shown how. Arup's use of facilitated workshops to achieve this, as an adjunct to their ostensible business purpose of sharing knowledge and developing new ideas about a topic of wide interest, is exemplary, and could be replicated anywhere.

Chapter Eighteen
Case Study: Broadway Malyan

An architectural practice can grow to the size of Broadway Malyan – in 2003, the fourth largest employer of architects and one of the top ten fee earners in the UK – only by providing good design and efficient delivery to a wide range of clients. Broadway Malyan recognised several years ago that continuing success would need a more systematic approach to learning and sharing knowledge, and it addressed the issue in an unusually insightful way. Rather than launch a generalised 'knowledge management' initiative, it decided to focus its effort on a specific business priority, and created a new post of business process facilitator with a brief to develop knowledge systems for supporting project delivery.

Practice profile (2005)

Staff: 430
Offices: 7 UK (10 across Europe)
Services: Architectural design, principally in education and research, workplace, residential, regeneration, health care and community, retail and leisure, urban design and masterplanning
Web: www.broadwaymalyan.com

Case study themes: Business Process software, Who's Who, contact database and induction process

---

Starting points

Broadway Malyan's new business process facilitator, Associate Adrian Burton, decided to tackle his brief by developing sophisticated, bespoke software that prompts, guides and helps job architects at every significant step in the life of a project, from job leader appointment to post-completion evaluation. Through a single interface, this 'Business Process' tool shows in detail what needs to be done; offers guidance, editable document templates and mail-merge facilities; creates an audit trail; and gives access to key project records. It also allows architects to search a 'Knowledge Forum' database for lessons learned in previous projects, and to add new knowledge as it arises.

Reflecting the practice's design-led, non-authoritarian culture, the tool does not dictate. Users can ignore most of the prompts if they wish, but there is every incentive not to: the system automates much of the administrative drudgery of job running, protects job leaders from procedural lapses, and frees time for creative design.

Business Process is an excellent example of knowledge codification, embodying key elements of the practice's collective expertise in project delivery, and making it available to even the most junior architect.

With that success behind them, Broadway Malyan was able to turn its attention to developing knowledge management systems to support the creative side of its work, while continuing to refine the support for process.

Contact with the Spreading the Word project made the practice realise that the most valuable knowledge is stored in people's heads, and is impossible to write down. This gave it a new objective: to connect people better and encourage them to talk more. As first steps, Adrian Burton and his team replaced an ineffective old skills database with a new Who's Who system, and redesigned their contact database and induction process.

Development of the Business Process system was well advanced when Broadway Malyan first became involved in Spreading the Word. It also had a variety of other knowledge management tools and processes in place, including:

- 'First generation' skills and contacts databases
- Project reviews and close-out meetings, informed by post-completion telephone interviews with clients conducted by job leaders
- Some specialist groups, such as a CAD user group
- Occasional workshops on specific topics, such as the Disability Discrimination Act
- A sustainability library, a collection of best practice details, and external resources such as Technical Indexes available through the intranet.

However, these had developed piecemeal, with no overall vision or strategy for knowledge management in mind, and person-to-person knowledge-sharing relied largely on informal networks and a cooperative culture. The skills database was little used, the contacts database was out of date, and there were no real communities of practice or systematic mentoring.

Business Process

The Business Process job management tool was envisaged as Broadway Malyan's core tool for sharing codified knowledge. In the basic structure of its task sequence, prompts and templates, the tool inherently embodies the practice's accumulated knowledge about administrative procedures – both those required for legal and contractual reasons, and those it has evolved to further quality, client relationships and efficiency. Broadway Malyan hoped that it would also gradually become a key dissemination route and point of access for a wide range of good practice information from internal and external sources, and the main place for staff to record lessons learned in projects.

Using a job management tool as the interface enables a proactive, just-in-time approach to knowledge. Where other approaches rely on staff to search, and perhaps overwhelm them with long hit lists, this makes it possible to offer them selected, relevant information just

Business Process tool screens

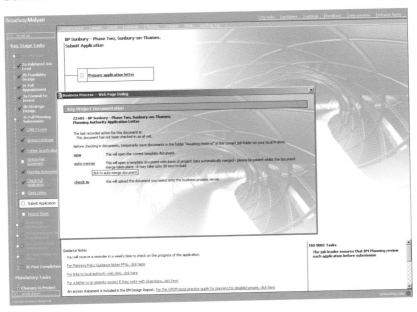

Figure 18.1 Broadway Malyan's Business Process system is navigated through a hierarchical list of job stages on the left of the main screen, and relevant guidance and templates can be accessed through links at the bottom of the screen. Here, a letter is about to be created in Microsoft Word by mail-merging a template and relevant project details, ready for editing.

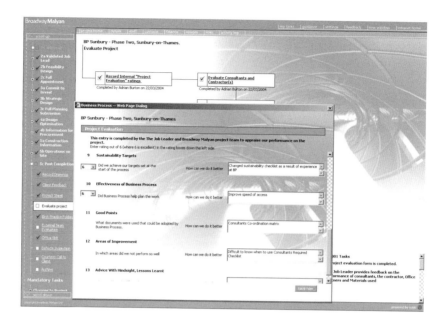

Figure 18.2 Job leaders are required to complete a project evaluation, and they can also evaluate consultants and contractors, and comment on suppliers and materials used. Guidance is also available for running hindsight review workshops.

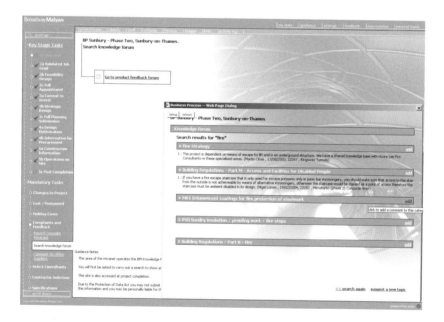

Figure 18.3 The Knowledge Forum database can be searched using a full-text index to find relevant lessons learned in previous projects.

when it is most likely to be useful. And they can be prompted to contribute new knowledge just when it is most likely to arise, as well. This suits an environment where day-to-day work is tightly focused on the job in hand, where the need for knowledge usually springs from a problem with the job, and where new knowledge arises largely as lessons learned on the job. Further, channelling guidance and lessons learned – in both directions – through a single tool that is in everyday use, at the time when people are most likely to be receptive, significantly reduces the psychological and practical barriers to both the recording and the reuse of knowledge.

Knowledge can age surprisingly fast, so the whole system is designed to evolve. Document templates, for example, are frequently updated in the light of comments from staff, conclusions from workshops, feedback from management meetings and project evaluations, and changes in legislation. Links to websites, calculation tools, internal guidance documents and examples of best practice are also kept under review. Having one central repository makes changes easy, and ensures that users can see only the latest versions.

To encourage the flow of new knowledge, the system automatically sends messages prompting job leaders to carry out project reviews, and there are simple electronic forms for recording client feedback and the results of project audits – with guidance on how to carry them out. Lessons learned become available to everyone as soon as they are entered into the system.

The tool helps management, too. For example, emails are sent automatically to let relevant directors know when projects pass key stages, and concentrating information such as audit results in a single database makes it easier to see trends.

Adrian Burton says the keys to success have been:

- Designing the tool to give users personal benefits – in this case, making it easier and quicker to generate project documentation and contacts lists than by conventional means
- Involving staff in the development, using questionnaires, focus groups and newsletters to get the benefit of their ideas and encourage them to feel ownership of the system
- Testing the business processes manually before software development starts, to ensure that they are pragmatic and flexible
- Keeping the process flexible, and allowing staff to override the software, so that it is seen as a help, not a straitjacket.

Bespoke software like this is expensive to develop: Broadway Malyan estimates the Business Process tool cost around £70 000. But it meets its needs in a way no commercial software could approach, and if it succeeds in raising the standard of project management throughout the practice to near the best, helps capture and disseminate lessons learned from experience, and enables architects to spend less time on dull administration and more on creative design, it will pay back its cost very quickly. In 2005 the signs were still good.

Who's Who

Participation in Spreading the Word convinced Broadway Malyan that it needed to do more to connect people and encourage them to share knowledge directly, person to person. To do this, people need to know who to talk to: easy in a practice of a dozen or two, where everyone knows everyone else, but difficult in one with 430 staff spread across ten offices. Some kind of networking tool is essential.

Broadway Malyan already had a skills database, but it was unhelpful and little used. It gave only an incomplete picture – information on CAD skills and training courses attended was kept elsewhere, for example – and the content was variable in quality.

The main features of a completely new system were finalised in 2005, and software development scheduled to start later that year. The new Who's Who was designed to make all the key people-related information available through one tool, and to serve several purposes:

- Making it easy for all staff to discover who knew what, and to make contact with each other
- Mechanising the generation of CVs for use in marketing documents
- Personnel management functions, including keeping training records and generating reports for use in staff appraisals: these would draw both on the Who's Who itself and on other sources such as time sheet records
- Providing an alternative point of access to the lessons learned recorded in the Knowledge Forum
- Acting as the hub of electronic communities of practice. Key topics were to have an appointed moderator, and staff would be able to post questions and sign up for email alerts when relevant contributions were made to the Knowledge Forum.

The intention was that training records would be entered by office training coordinators to ensure that descriptions were consistent, and when new records were entered the system would automatically email trainees to prompt them for feedback on courses.

Like the Business Process tool, Who's Who was expected to continue to evolve for some time after coming into use. It was not clear, for example, how skills could most usefully be recorded. Searching is easiest with predefined categories (but they are inflexible) and skill levels (but they may be applied inconsistently), whereas free-form descriptions are richer, but variations in terminology ('housing' and 'residential', for example) complicate searching.

Experience in other industries has shown that details and psychology can make all the difference to the success of knowledge management initiatives, and the system was designed to use both carrots and sticks. Simplification of tedious administration was the main carrot, and the default entry for skills was one of the sticks: it declared 'I have no skills to offer'.

Contact database

It is as important to know who you know outside a practice as inside it. Until recently, Broadway Malyan's records of external contacts were in much the same state as their skills records: often out of date, and divided between several unconnected systems – a main contacts database, separate marketing and event invitation mailing lists, and numerous private lists in Microsoft Outlook. These were replaced in 2005 by a new database designed to provide a single repository for contact data, link it richly to other related data, and create a system that is both easier to maintain and more useful.

Contact database screens

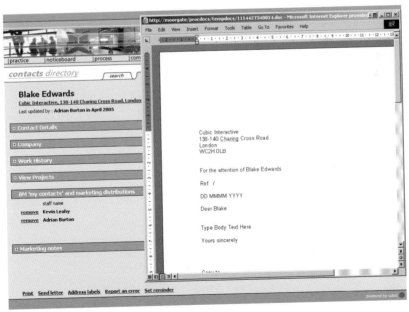

Figure 18.4 Broadway Malyan's contact database: letters and address labels can be created directly from the database by adapting standard models.

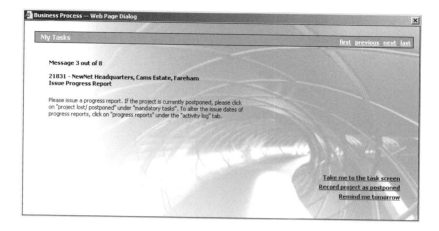

Figure 18.5 Prompts – here, a reminder about a project review – are generated by combining information from the contact database and the Business Process tool.

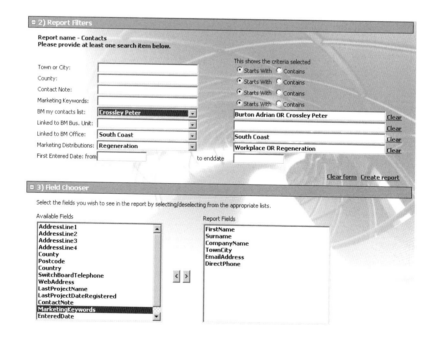

Figure 18.6 A 'report generator' tool allows bespoke contact lists to be created in Microsoft Excel.

There are numerous proprietary contact databases (not least Microsoft Outlook itself), and many of these incorporate powerful and useful facilities. Broadway Malyan examined several, but concluded that bespoke software tailored to its particular requirements and IT environment would be a better buy, despite its probable higher cost. The assessors found that commercial software was heavily biased towards salesmen, emphasising activities such as recording conversations and following up leads that are peripheral in design practice, at the expense of simplicity and ease of use. It would be more difficult (and in some cases impossible) to integrate with the practice's other systems and databases, too.

The system Broadway Malyan developed is user-friendly and tightly integrated with the practice's Business Process tool, and with staff and other databases. New and updated information entered into contact fields in any of them immediately becomes available to all, greatly reducing the effort of keeping information current, and avoiding information conflicts.

Integration also allows the database to keep staff informed about:

- What projects contacts have been associated with
- With which consultants the practice has framework agreements
- Who knows who.

Finally, the system is designed to go beyond its basic function of providing contact and contact-related information and mechanise a number of tedious administrative tasks. Additional services include:

- Mail-merged letters, address labels, and 'remind me' messages
- Project contact lists
- Mailing and invitation lists for specific purposes
- Links to related information on the Web, such as route finders, directory enquiries and street maps.

Facilities such as these give staff immediate, visible benefits that encourage them to help keep information current.

Induction process

Broadway Malyan does not have a systematic mentoring system, but it has revised its staff induction process to give some of the same benefits.

It is easy to forget the amount of knowledge that new entrants need to absorb to learn 'how we do things around here'. Packing it all into one or two days of briefing can easily overload them. Broadway Malyan now:

- Splits induction briefing into small, digestible packages
- Allows people to learn at their own pace, on a flexible timetable
- Uses a variety of media to suit different areas of work, including the practice's public website, its intranet and animated software demonstrations, as well as face-to-face briefing
- Uses electronic checklists to ensure that entrants meet all the people they need to, and are briefed on all the topics they need to know about (without being burdened by information irrelevant to their jobs). Where appropriate, the checklist is linked directly to electronic briefing material
- Maintains an audit trail showing when people are satisfied that they have been adequately briefed on each topic
- Includes a meeting at the end of the process to review any training needs.

Last, but not least, the process enables new entrants to make a range of personal contacts spread over enough time for them to be individually memorable, and creates a starting point for building a personal network and sharing knowledge in future.

Together with the new databases and tools it is developing, Broadway Malyan hopes that the new induction process will mean that, in the future, the knowledge in the heads of its staff becomes more and more of a shared asset.

Commentary

Broadway Malyan's story is incomplete, because the Spreading the Word project finished before most of the new tools and processes had been launched. The specifications for the various tools are impressive, particularly in the evidently careful consideration given to users' needs and point of view, but no information is available about their success in practice.

Chapter Nineteen
Case Study: Buro Happold

Founded in Bath in 1976 by the late Professor Sir 'Ted' Happold, Buro Happold is one of the UK's leading multidisciplinary engineering consultancies. Set up initially as a specialist structural engineering practice, the firm has grown organically over the years to provide a variety of other services.

The partners have long believed that clients are best served by an integrated, multidisciplinary service, and they have made this approach a central tenet of the company's work philosophy. Nevertheless, Buro Happold's offices have traditionally been organised by engineering discipline, with structural engineers and building services engineers working in organisationally and physically separate groups. In a recent review the partners concluded that this should change: to improve multidisciplinary working, engineers should work in future in 'integrated business groups' (IBGs) made up of staff from a range of disciplines sitting together.

As refurbishments become due, a new workspace design is being rolled out to support the new organisational structure and actively encourage interdisciplinary collaboration and knowledge-sharing in general. The new workspaces are also more space-efficient, so they will help accommodate growing staff numbers.

Since the case study, Buro Happold has been through a period of soul-searching about its management of knowledge. This has led to a decision to focus on connecting people and not to try to develop documentary knowledge resources. A new IT-based 'Knowledge Directory' embodying this philosophy has recently been launched.

Practice profile (2005)

Staff: 772 UK staff (952 worldwide)
Offices: 6 UK offices (14 worldwide)
Services: Building engineering, infrastructure, transport and urban
           development, environmental consultancy, and project
           management
Web: www.burohappold.com

Case study theme: workplace design

Starting points

The London office was made the prototype for the new workspace design. 17 Newman Street had become an uninspiring place, dominated by filing and desktop computers. Visitors could have been excused for not realising that it was an engineering design office – the only clues were a few framed pictures of completed projects.

To help realise its vision for the new workspaces, Buro Happold called in specialist design consultancy DEGW and consulted other companies the partners admired, such as product designers IDEO. It set up a project team to manage the process, and they visited various furniture showrooms to discover how far the space efficiency and flexibility of new desk systems could improve on Newman Street's existing furniture. After a series of initial investigations, refurbishment started in early 2003 with a small area, to test the new design. Reactions to this were encouraging, and the rest of the office was completed about a year later.

Buro Happold staff were consulted, too. Their wish list was simple: storage, daylight, and a working computer. As long as they had these, they expected to be happy. But the refurbishment project team wanted the new design to do more than simply meet the most basic needs: they wanted to change working habits for the better. In particular, when traditional drawing boards were abandoned in favour of CAD, it had been noticed that the discussion of design dwindled: it is next to impossible at a computer screen on a normal desk. Buro Happold wanted the new workspace design to bring discussion back.

In addition to being asked what they wanted, staff were observed to see how they worked and interacted. This gave the project team many insights. They found, for example, that desks were unoccupied for long stretches of time; that designers had nowhere to lay out drawings; and that staff did not get up to speak to people sitting more than a few metres away, but relied on email or phone.

The prototype

At this stage designers from Design Engine Architects became involved, and they eventually took the design forward to prototype and final design. They considered the wider office environment as well as the individual workspace, and they looked at the relationship between the various activities an office has to accommodate and the spaces where they can take place. As part of the process, they carried out a series of studies to analyse how desk space was currently being used, and how it could be used in future if additional, shared spaces were provided to allow selected activities to be shifted away from the individual desk.

A prototype workspace was set up to their design on the ground floor of the office. This had a number of key innovations:

- Workbenches at two heights: 725 mm (the conventional height) and 1050 mm. The higher benches were designed to bring the heads of seated people and standing colleagues to the same level, making it possible for the first time to hold useful discussions around a computer monitor, and facilitating conversation in general
- Flat screens rear-mounted on movable brackets and posts, to save desk space
- CPU racks at the ends of the benches, decluttering desks and facilitating maintenance
- Personal storage units on wheels, to make it easy for people to change seats.

Carpet was replaced with a hard linoleum floor, and white walls with blocks of bright accent colours replaced an off-white, bland colour scheme.

Design Engine also designed a new raised, open meeting space at the back of the ground floor office, recognising that the area was a focal point for anyone entering the main floor area. This wall had previously been covered with shelves of filing, and these were replaced with metal ceiling tiles and sliding whiteboards to enable designs to be pinned up and discussed.

Sixteen staff were invited to use the new desk spaces, and everyone in the office to use the new meeting space. After a few weeks, staff were asked for their comments. They liked the new desk spaces and flexible layout tables. They approved of the better shared spaces, and liked the under-desk storage units, the new, simpler colour scheme, and the whiteboards and metal walls for pinning up drawings. They found the high workbenches challenging, but they recognised the opportunity these gave to collaborate and interact more freely.

The final design

Design Engine took the best ideas from the prototype and used these to inform the design of the rest of the office, including workspaces, social spaces, the library, kitchen/dining area and meeting rooms.

The final design incorporates several new breakout and meeting room spaces, varying in formality from the table area designed to encourage spontaneous gatherings among project teams, to more formal conference rooms. All the meeting areas are arranged around the edge of the floor plan, next to the windows and stairwells, surrounding the centrally placed workbenches.

Inspired by bookshops that have thriving coffee shops where people meet to talk, the dining area and library have been combined – for many staff, lunch hour is the only convenient time to 'browse'. The dining area can also be used as a formal or informal meeting space throughout the day.

Evolution of a workplace: Buro Happold

Figure 19.1 Buro Happold's offices before the redesign.

Figure 19.2 After the redesign: decluttered and interactive.

1. Work benches 1050mm high

2. New chairs to match the high benches

3. Flat computer screens rear-mounted on movable posts and brackets

4. Computer racks at the ends of benches

5. Personal storage trolleys

6. Walls covered with perforated metal tiles to use for displaying drawings (attached with magnets) and as a projection screen

7. Sliding white boards with metal backs for sketching and displaying drawings

8. Meeting spaces

9. Hard linoleum instead of carpet

10. Layout tables for drawings (out of shot)

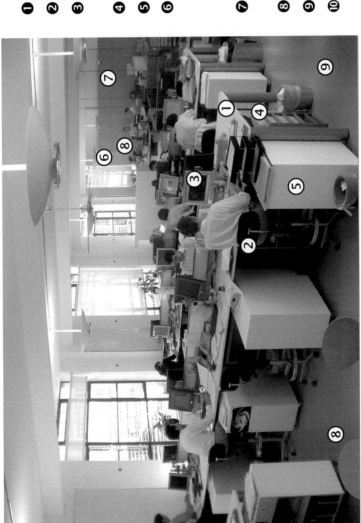

Figure 19.3 The main changes in Buro Happold's office.

Breakout areas and hot desks have been provided on every floor. In the past, hot desks were separate from the workspaces, typically near the front door; placing them within the work areas has increased the opportunities for interaction and knowledge-sharing between local and visiting staff.

The result is an office in which personal workspaces are smaller than they used to be, but they are less cluttered, and there are many more shared surfaces and spaces to use. This rebalancing between personal and shared space deliberately favours collaboration and ad hoc conversation, and in the long term it is expected to make a real difference to knowledge-sharing.

Assessing the results

Buro Happold canvassed staff opinion on the original workspace before the prototype area was occupied, to provide a benchmark against which the new design could be assessed, and on the new arrangements after the whole refurbishment had been completed.

The refurbishment project team developed a bespoke questionnaire based on the Office Productivity Network Survey, with some ideas brought in from a survey designed by consultancy Building Use Studies. Both the 'before' and 'after' surveys asked respondents to assess their satisfaction with office facilities, with questions on the space, furnishings and equipment. In the 'after' survey, respondents were also asked how well they thought the new office layout supported a range of specific tasks and activities such as collaboration, quiet concentration and creative work. Five-point response scales were used for all the questions. The questionnaires were distributed to different samples of 36 employees; 24 completed the first survey and 14 the second.

Among the 'public' spaces, the new breakout, conference and café areas proved popular – the café area even brings people together from different floors. On the other hand, there was widespread dissatisfaction with the areas provided for reading and quiet study: this appears to be largely an acoustic problem, and should be relatively easy to fix. Research has shown that noise is a central issue in the success of open-plan offices, so it is not surprising that Buro Happold's new fit-out needs some fine-tuning in this respect.

The before and after assessments of personal workspaces show no significant changes in satisfaction with desk space, storage or chairs. Reactions to the high desks and chairs are mixed, as they were in the pilot trial. Mobile staff and managers whose work is communication-based are relatively unconcerned about the height, but it is more controversial with technical staff who spend long periods at their desks. One clear message is that though the new design provides more space than the old office to spread out drawings and plans, staff would like even more.

Perceived improvements in collaboration, creativity, concentration, minimising errors at work and meeting deadlines are all smaller than Buro Happold hoped. This is not surprising. The organisational changes that took place at the same time will have had a much larger influence on behaviours such as these than the office redesign, and the unsettling effect that organisational change always has will inevitably have coloured reactions to the new office. It will be interesting to see how staff assess it in six to nine months' time, when they have become used to working in integrated business groups. Anecdotal evidence suggests that perceptions are already becoming more clearly positive.

As well as giving valuable feedback on the new office, the surveys have had the incidental benefit of increasing awareness of the effect that the working environment has on staff performance and well-being. Respondents valued an opportunity to express their opinions and influence the design of their workspace, and they were keen to hear what results emerged from the survey.

Having taken first steps towards developing both a workspace design that encourages knowledge-sharing and a systematic way to assess the effect of design features, Buro Happold intend to continue making and monitoring changes until the vision becomes a reality.

Commentary

Designing workspaces to meet the complex needs of an organisation such as a multidisciplinary consultancy is not easy. Buro Happold's step-by-step approach – careful design, a small pilot, assessment, a larger trial, further assessment and a period of fine-tuning before large-scale roll-out – shows how it should be done. 17 Newman Street has not solved all the problems, but it undoubtedly represents a significant step in the evolution of an office fit for the knowledge age. At the same time, it allowed density to be increased and solved a growing space problem without alienating staff.

Since Spreading the Word, Buro Happold has turned its attention to other aspects of knowledge management. With the help of an independent consultant, it reviewed a variety of options before concluding that 'talking to each other' was so much a part of its culture that it should make connecting people the core of its knowledge strategy, and not try to develop documentary knowledge resources.

Launched in 2006–7, its new 'Knowledge Directory' aims to 'connect groups with groups, people with people, and skills with projects'. The main navigation tool is an expandable tree view organised at the top level in sectors, subsectors, components, 'ingredients' (sources of knowledge on key aspects of practice, such as CDM, construction sequencing, and health and safety), the geographic location of projects, Buro Happold offices and groups, and disciplines and people. These top-level headings are each subdivided into one or more further levels. The sectors section, for example, expands to 12 subsections (from commercial offices, commercial residential and cultural and civic buildings to retail and sport, leisure and events), and each of these divides further. Each sector and subsector page contains lists of links to relevant projects, locations, capability statements, managers and experts, and to miscellaneous documents; some also have an introductory text. Following links eventually leads to individual 'person pages' and 'project pages' that contain respectively information about people's location, responsibilities, qualifications and project experience, and a variety of basic factual information about projects, images and marketing material.

As it has finally been implemented, the Knowledge Directory departs considerably from the declared intention to focus on person-to-person contact and avoid codifying knowledge – an approach that, taken to extremes, is unlikely to meet any consultancy practice's needs. The Directory is much more than a networking tool; it is also a project directory, and potentially gives access to a substantial amount of other codified knowledge in the introductory text on many of the pages and through links to separate documents. It is too early to judge how well the mixture of networking tool and knowledge base works for Buro Happold; it will be interesting to see which way the balance between the two aspects tips in the future.

Chapter Twenty
Case Study:
Edward Cullinan Architects

For Edward Cullinan Architects (ECA), participation in the Spreading the Word project coincided with a period of soul-searching about the future. The practice was growing, and was about to reach the notoriously difficult size where an organisation can no longer work as a 'family' in one space, senior members find themselves increasingly stretched by the demands of managing the business, and specialist managers are difficult to afford. With founder Ted Cullinan in his 70s, the practice also had to start planning for a reduction in his involvement.

ECA found that thinking about knowledge management provided a helpful framework for thinking about the whole future of the practice. It found, too, that many of the working practices that had developed intuitively over the years had been silently helping them to learn and share their knowledge. ECA had in fact been managing its knowledge very effectively – but it was rapidly approaching a tipping point beyond which it would need a more conscious and systematic approach. This is director Colin Rice's description of how it started to develop a new knowledge strategy to meet the needs of a bigger, changing practice in an increasingly demanding market.

Practice profile (2005)

Staff: 28 in June 2003, 39 in March 2005
Office: 1 (London)
Services: Architectural design, particularly masterplanning and mixed-use urban regeneration, visitors' centres, education and housing
Web: www.edwardcullinanarchitects.com

Case study theme: knowledge strategy

Edward Cullinan Architects is the smallest practice to participate in Spreading the Word. It was founded in 1965 by Ted Cullinan with a powerful ideal of a cooperative practice focused on designing and making beautiful buildings that respond gracefully to their users' needs and their physical context. Success in achieving this vision can be measured both in the steady flow of awards, and in the high esteem in which Ted Cullinan is held by the profession.

Our principal areas of work have evolved with changing economic circumstances, and will continue to do so. A strong lineage of visitors' centres to historic places, from Fountains Abbey, through Stonehenge, Archaeolink at Oyne, and the Weald and Downland Open Air Museum, now continues with projects for Petra in Jordan and the Cambridge and Edinburgh botanic gardens. About 40% of our work is in mixed-use urban regeneration, encompassing masterplanning and housing. Education has been a strong strand of work since 1990, with notable projects for Cambridge University, Warwick University, and more recently Sandwell, and Greenwich Millennium School.

Success has attracted larger projects, which in turn has created pressure to grow. Participation in Spreading the Word coincided with a period of particularly rapid growth, from 28 architects and support staff in August 2004 to 39 at the time of writing in March 2005. This challenged many of the processes and arrangements that could be seen as 'natural' knowledge management for a small practice, and it made us think hard about fundamentals, rather than merely adjust some of the dials. As a design practice whose asset is creative ideas generated and realised by individuals working together, we came to recognise that we needed to develop a coherent strategy for managing knowledge as we become a larger and increasingly different organisation.

A further challenge comes from rapid change in the outside world. Communications technology, IT, the regulatory framework and the current revolution in the construction industry as a whole all make it impossible even to attempt to stand still.

Growth is changing our culture, too. Historically it combined a cooperative management structure with strong leadership from Ted Cullinan in determining the direction of design. This put a natural limit to growth, in that there are only so many projects in which one person can be intimately involved.

A culture in which everyone, including the younger members, is positively encouraged to have their say and participate makes a strong foundation for knowledge management. We had long-standing techniques for knowledge-sharing in place, although not labelled as such. In planning for change we were determined not to abandon the strengths of our culture.

The practice has never been 'commercial', in the sense that running the business has always been secondary to the goal of making great architecture. But in an increasingly competitive world, keeping up with the game in terms of processes is crucial. In terms of Treacy and Wiersema's 'value disciplines', we have firmly set product excellence as our primary objective.

Starting points

On reflection, we were able to recognise a number of our established working practices as 'knowledge management'.

Workspace
We work in a single open-plan studio with an understanding that everyone should keep ears and eyes open to conversation and work on the drawing boards. Knowledge-sharing is helped by the key players having naturally loud voices, and by everyone moving regularly with the ebb and flow of projects. This used to be effective in creating a cohesive group, but as numbers have crept up from about 25 to 39 in the same space, and drawing boards have been replaced by computers, it has become less so. At present the zone of real day-to-day influence is probably only about a third to a half of the office.

When, for a time, the office was split between two floors the effect was to create a 'them and us' division. This has remained a warning and an influence on our thoughts about workplace design.

Friday lunch
Everyone takes turns to prepare a sit-down Friday lunch, at the end of which there is a short meeting to deal with housekeeping, details of new jobs, forthcoming CPD events and seminars. People are encouraged not to plan external meetings at this time so that there is generally high attendance.

Weekly CPD sessions
A weekly CPD slot on Wednesday has been running for about ten years. Organised by a rotating group of three people, this combines:

- In-house presentations of current projects
- Presentations by firms on particular materials or products of interest to the office
- In-house seminars by 'topic champions'
- Feedback from people who have been to outside seminars or conferences.

Topics are selected in the light of an annual review of individual personal development plans. Now that CPD has become mandatory for RIBA members there is a good turnout most weeks (average CPD is 40.5 hours per annum, in addition to 45 hours' training).

Figure 20.1 Friday lunch at Edward Cullinan Architects.

Office handbook

The scope of this is wide, covering:

- Practice philosophy
- Staff welfare
- Facilities
- Day-to-day running
- PR and marketing
- Project management
- CAD procedures
- IT procedures, such as use of Microsoft Outlook and PowerPoint.

The handbook began as a dog-eared loose-leaf file bristling with Post-It notes, but it is now electronic, and quickly accessible through the office intranet, and this has increased its usage and authority.

Database

This contains key project information and the office address book. It used to be a home-made FileMaker Pro database, but that was slow and awkward to use, and over the course of the past year it has been made more accessible by a web browser interface.

Project reviews
We hold project reviews at each RIBA work stage. These are of two types:

- Design reviews on the architecture. A group of reviewers is assigned to stay with each project throughout, but review sessions are open to all, and they usually attract a handful of 'outsiders'.
- Project reviews, dealing with the process side of the projects.

Project reviews are all done by the same person – the quality manager required by ISO 9001 – so that divergence from office practice can be corrected, and new ideas fed back.

Stage reports
Stage reports are standard practice for all jobs, and available to all to read. These do get read, and they inform reports on subsequent jobs.

Lessons learned
We have talked for years about improving our practice by building up a library of annotated as-built drawings and specifications. However, this has not been systematically carried through.

Topic champions
Everyone in the practice is designated either a 'champion' or a 'supporter' of a particular field of interest, such as higher education, offices, access, or ethics. Champions are supposed to take responsibility for keeping abreast of developments and ideas in their field, and go to relevant conferences and seminars, but again this has not really happened: people tend to be champions of the subject of their last job rather than their current one.

Mentoring
Younger members of the practice are each allocated an 'uncle' or 'aunt' until they have taken their Part 3 exams.

Knowledge strategy

On paper, our single studio workspace, project reviews, topic champions and various other practices together make a good foundation for knowledge management. In reality, we relied until recently very largely on the informal elements – our social workspace and Friday lunches –

which are precisely the ones most at risk from growth and organisational change. Over the past year some of the more formal elements, such as the office handbook and project reviews, have started to make useful contributions, but others remain good intentions. In reviewing our future, we decided that we needed to develop a coherent and practical knowledge strategy – not just a planning document, but an ethos that could underpin all the changes to the practice.

Involvement in Spreading the Word showed us that there is no single fix: knowledge management needs to inform all our professional work and management. Everyone needs to keep abreast of constantly changing regulations and other documented information. Tacit knowledge about design and about the culture of the practice – 'how we do things round here' – needs to be passed on to a flow of new staff.

A good strategy needs to include a systematic approach to learning and to both basic patterns of knowledge-sharing – direct transfer through person-to-person contact, and indirect transfer through codified knowledge.

With this in mind we have looked at a range of tools and techniques and considered what they could contribute, and how they could be implemented effectively, in the context of our practice.

Workplace design

We moved into our present building in 1991 when there were just 20 of us, most work was done on drawing boards, and there were just a few computers between us. Like desert nomads we brought our office layout with us from our previous building: perimeter layout desks, with drawing boards arranged to give face-to-face contact. It seemed to work. As computers replaced drawing boards, and numbers grew, this evolved into a bay arrangement. Capacity has grown as flat screens have replaced large monitors, and A3 files have replaced plan chests and A0 drawing clips. Where people sit is decided through a process of bimonthly resource reviews followed by a review of 'who goes where', moving people around so that team members are co-located.

But inflexibility has made the bay system begin to creak at the joints, and having reached capacity in our present studio – and having been influenced by discussions in Spreading the Word workshops – we have drawn up a new plan for the office. Our overriding goal remains to continue the ethos of all working together in a large studio, but the knowledge management perspective has added some subtle nuances.

An audit of the way we use codified knowledge showed that the technical library has been largely superseded by electronic sources, so the space can be released for other uses. In future, we plan to get information on materials and components entirely from a combination of:

- External electronic sources such as Technical Indexes, Google and the RIBA product selector accessed from the NBS pane
- A categorised database of information gained in our projects and research.

Features of the new plan that we hope will support and encourage knowledge-sharing include:

- 'Magnets' such as printers at the ends of the office rather than in the 'logical' position at the centre, to encourage people to wander
- Eye-to-eye contact across the work tables, which had been lost in the bay structure
- All filing put on a service wall to make it easier to access (together with a reduction in paper filing to free up space)
- Wall-mounted pin-up space in place of shelves, to restore the visibility of work in progress and the opportunities for discussion that we lost when CAD came in.

Induction
As the practice has grown, a more formal induction process has been introduced, and a member of the management team now works through a checklist with each newcomer.

Evolution of a workplace: Edward Cullinan Architects

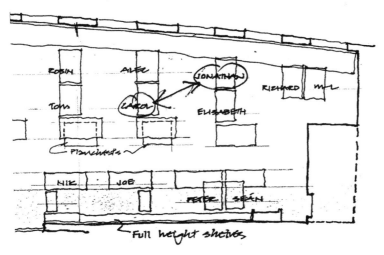

Figure 20.2 ECA's office in 1991, with drawing boards, face-to-face contact, shelves along one wall, and ample space.

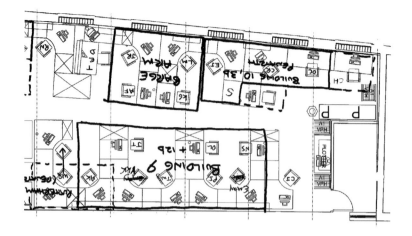

Figure 20.3 Today: bays, computers, teams in relatively isolated groups, crowded.

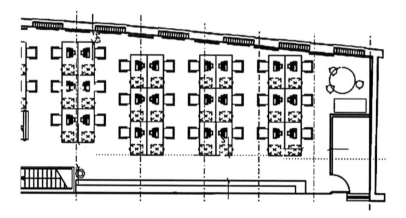

Figure 20.4 ECA's plan for their future office. Work tables (seating whole teams or parts) with eye contact across them and free circulation around them, 'magnet' and breakout spaces to give reasons to wander, support staff sitting with architects, pin-up space on the wall, and fewer paper files.

Networking

In the past the ease of simply asking around has made a skills database or 'Yellow Pages' staff directory largely superfluous, and we have never succeeded in developing a useful system. We plan to keep a watching brief on this.

Job running and project reviews

We have been conscious for a long time that we need to standardise job-running processes, and to learn more from our experience. IT has made this both more practicable and, as the day-to-day running of the practice has become less visible, more necessary.

Until five years ago, individual job architects set up their files as they wished. This made it difficult to switch between projects and find information easily. A common filing system was rolled out in 2000, and we now have an electronic system in which the electronic folders mirror the paper files.

A comprehensive office handbook has been developed over the past eight years. In parallel, the IT manager has set out standard CAD and IT procedures in CAD and IT manuals. These remain evolving documents. Usage has increased since they were converted from hard copy into an electronic guide on the intranet, supported by a good search facility, and they formed the basis of achieving ISO 9001 accreditation in July 2004.

The hurdle that had to be crossed to achieve accreditation was less in having the procedures in place than in auditing their usage, and in using the audit process to feed improvements back into the system. The project review process is one vehicle for closing this loop. Reviews take place at each RIBA work stage, involve the whole team, and follow a set agenda. One of the reviewer's roles is to tease out whether communications in the team are working: do people all understand the big project vision, and do they know what they have to do and by when?

Project reviews, only sporadic a year ago, have now become a regular part of a project's evolution. They have changed from something we knew we should do to something we *do* do, with tangible benefits in improved communication within teams and across the office.

Hindsight reviews

After several experiments, the culture of carrying out hindsight reviews is spreading.

A recent hindsight review of a competition (carried out before we knew we had been successful) illustrates the approach we use. As many team members as possible attended, and it was chaired by the director in charge. Following the After Action Review format, we discussed what happened step by step, whether it was good or bad, and how it could have been done better. The conclusions were recorded in two columns, one 'facts' and the other commentary – a format we find helpful because it adds the 'wisdom' overlay in a clear way – and the report was posted on the wiki (discussed below).

Will anyone ever read it? Possibly not, but the benefits are as much in the process as in the product. It is, for example, an opportunity to dispel myths that rapidly arise about a project: everyone has their own perspective on what and why things happened, and without formal review often only the loudest voice is heard. And involving the younger members in the process has a direct benefit in morale: 'This is an office where my views are actively sought.'

This particular review yielded several useful lessons. It showed us, for example, that a movie record would be valuable for initial surveys of a large site, and that the competition team would have benefited from stronger 3D CAD skills. As a direct result, we purchased an easy-to-learn package that seemed ideal for early-stage work (SketchUp), tested it on another job (to great effect), and we have since bought five licences.

We have also started carrying out hindsight reviews on practice presentations for competition interviews, given to the whole office in a Wednesday CPD slot. These have multiple benefits: they give practice at giving presentations, they provide an opportunity for others to comment on the content, and they help spread understanding of the latest ideas.

We have found hindsight reviews very effective – not least because they combine face-to-face knowledge-sharing with codification.

Wiki knowledge base

We are at an early stage in setting up a wiki, accessed through the practice intranet. Open source wiki software makes an attractively affordable alternative to commercial or bespoke software for a knowledge base, particularly in a small practice – without sacrificing power.

The basis of our system was set up by the IT manager using the TWiki package,[1] and introduced to the whole office in a Wednesday seminar slot.

We started with a jump-in-and-splash approach to structure, but we have since rethought this; one of the beauties of wikis is that they are so easily adjusted. The new structure has basic sections following the structure of the management teams in the office. The team leaders are thus the natural 'moderators' of their sections. Topic champions now have somewhere to record the material they collect. We hope it will also give them more incentive to keep up to date and play their intended role.

It is taking some time to decide what sort of material to include in the knowledge base. Should it, for instance, include 'passive' material – things such as reference material and reports, which are unchanging and could just as easily be made accessible through the intranet – or only 'active' material to which insights and comments can be added in a more organic way? One point on which we are clear is that a wiki is an ideal home for all the little scraps of information that people

---

[1] TWiki information and multiplatform software is available from http://twiki.org.

have historically accumulated in private collections, but which could usefully be shared with everyone else; it should take no more time to put information like this in the wiki, where it can benefit the whole practice.

It is much too early to judge whether the wiki will be a success, but about 15% of people in the practice have contributed material so far and, after a slow start, usage is increasing.

Conclusions

We have learned a number of important lessons from our participation in Spreading the Word and our thinking about knowledge strategy and the future of ECA. The most fundamental of these are that:

- Knowledge management has to be an ethos underlying everything a practice does.
- Solutions must, therefore, be continually evolving.
- A knowledge strategy must have a broad front: there is no single answer.
- The first priority is to promote face-to-face contact for learning and knowledge transfer.
- But it is important, too, to codify knowledge where possible and to be able to store and find codified knowledge easily.

Commentary

ECA's experience is typical of well-run, small professional services organisations. Knowledge flows naturally and freely when a small, stable group of professional peers work in a single space, the management style is non-authoritarian, and there is mutual respect and trust. A knowledge ecology evolves in which deliberate 'knowledge management' is almost superfluous. Rapid growth and larger numbers, however, can both disrupt it, potentially to breakdown. ECA was faced with both.

When numbers increase by 40% in under two years, as they did at ECA, long-serving staff (who are inevitably even busier than usual) simply do not have time for the one-to-one contact that is the mainspring of natural knowledge-sharing – just when it is most needed to indoctrinate the newcomers into the practice's ethos, standards and ways of working.

Larger numbers have wider-ranging effects. As ECA noticed, when they rise much above 20 they lead to a reduction in overhearing, and when they force staff to be divided between separate rooms and floors 'them' and 'us' groups can start to develop, with decreasing contact between them. At the same time, it becomes increasingly hard for everybody to play a full part in meetings such as project reviews and ECA's Friday lunches, and impossible for unique physical resources such as ECA's hard-copy office handbook to be shared effectively. Effects like these escalate dramatically when practices open new offices – something ECA has yet to face.

The kind of learning needed to extend already high levels of professional competence comes less naturally – particularly in architecture, a profession that has no tradition of looking back, and in which the urge is always to move on to the next job. As we saw in Chapter 2, simply doing the job has only limited effect; achieving exceptional performance requires reflective thought and continuing challenge. It is not surprising, therefore, that the learning practices that evolved at ECA fell short of the excellence of their knowledge-sharing.

A high rate of growth and larger numbers both necessitate a more conscious and managed approach to learning and knowledge-sharing, as ECA recognised. The increasing technological sophistication of buildings and ever-more demanding regulations and markets add further urgency. The practice took an exemplary approach to developing a knowledge strategy, beginning with a review of its aspirations for the business, thoughtful review of its existing informal practices, and evolutionary development to correct its weaknesses (in project review, for example), build on its strengths, and make it more systematic. Wisely, the technology was kept low-key, and a software framework was chosen that was capable of evolving with the practice's changing needs in the future. A wiki can, for example, not only support any likely future requirements for the knowledge base, but also provide a ready-made framework for a networking directory if and when it comes to seem necessary.

That was in 2005. In the subsequent two years, ECA's rate of growth has slowed to a more manageable 10–15% a year, and the practice has succeeded in retaining its basic character and ethos. Colin Rice has taken on a new full-time practice management role, and with it the opportunity to spend more time on improving the practice's organisational learning. The strategy developed during the Spreading the Word project has been implemented to a considerable degree, and the ideas behind it continue to inform its development. The proven elements in the knowledge strategy have largely continued unchanged, and most of the newer ones have continued to establish themselves and evolve – sometimes in unexpected directions.

The *office refit* has yet to happen. ECA developed a scheme to redevelop the site of their office, building flats on part of it to fund a new office alongside, but this has been held up by planning and other obstacles. The directors have taken the opportunity of the delay to visit several other practices that have already refitted their offices along the lines they intended, and this has reinforced their conviction that it is the way they want to go.

Induction has been further strengthened. New recruits are introduced to a range of key staff, and given time to sit down with them and be introduced to the practice's systems and procedures. Any who are not familiar with the latest version of the CAD software ECA uses are sent on two- or three-day courses.

A *Yellow Pages* database has been set up, with details of everyone's qualifications, experience, projects, interests and professional activities outside the practice, such as teaching. This has proved worthwhile – overturning the judgement that Yellow Pages were superfluous in a practice of around 40. One feature that has not proved a success is a section recording people's IT skills: these are widely enough spread to make it equally effective just to ask around, and easier. ECA is still grappling with the problem of persuading staff to keep their entries up to date.

A new *project database* has also been created, and drawings and specs from past projects are gradually being added to it as pdfs. Using pdfs allows comments to be added freely, and goes some way towards realising ECA's long-held ambition to have a library of annotated as-built drawings as a repository for lessons learned.

Project reviews have become routine, and more systematic. *Hindsight reviews* have also become a regular part of 'how we do things around here' after competitions and the most interesting projects. The two-column fact and commentary layout adopted from learning histories has proved its worth as a mental discipline and a good way of recording lessons learned.

The *wiki knowledge base* has been the one clear failure. It has recently been abandoned, and replaced by a straightforward database and keyword system, which acts as a portal to web links and separate document files. These are largely pdfs, so they can be searched internally and can be (and are) annotated. The failure of the wiki appears to have been due largely to lack of a sufficiently well thought-through and enforced structure: this left people uncertain where they should put, or look for, information, and the content eventually became (in Colin Rice's words) 'too chaotic to use'. It is too soon for definitive judgement, but so far the more rigid constraints of a conventional database and keywords appear to suit ECA users better. It will be interesting to see whether this remains the case in another two years.

Chapter Twenty-One
Case Study:
Feilden Clegg Bradley

Founded in Bath nearly 30 years ago, Feilden Clegg Bradley Archi-tects[1] (FCBa) has always had innovation, learning and knowledge-sharing at the heart of its practice philosophy. With its special interest in environmentally conscious architecture it had to; few people knew anything about it in the late 1970s. Since then it has continued to develop its expertise, both through practice and through ongoing involvement in research on energy, materials and the performance of buildings in use. In the past few years it has become recognised as one of the country's leading exponents of sustainable design, and its boundary-pushing architecture has won a string of awards, culminat-ing in BD Architect of the Year 2004 – with awards for arts/culture buildings, public housing and private housing – the Queen's Award for Sustainable Development 2003, the Civic Trust Sustainability Award 2003, Architectural Practice of the Year 2003, and numerous accolades for individual schemes from the Civic Trust, RIBA, Housing Design, HomeBuilder and others. Being a learning organisation has paid off handsomely.

With recognition came expansion. The size of FCBa's largest proj-ects grew from around £10 million to over £70 million between 2001 and 2005, and staff numbers more than doubled to a total of 20 part-ners and around 95 staff. Having more staff working on more and bigger projects meant more opportunities to learn, but at the same time made it increasingly difficult to be confident that lessons were still being learned and new knowledge was still being widely shared. After the opening of a second office in London it became clear that relying on personal initiative was no longer good enough, and a more systematic approach was needed.

Since then, partner and practice manager Chris Askew and partner Ian Taylor have taken a series of knowledge management initiatives, including:

---

[1] Now Feilden Clegg Bradley Studios.

- *Hindsight reviews, to improve learning from projects and the sharing of lessons learned within design teams*
- *Developing an embryonic skills database into a fully fledged networking directory to help to connect people and facilitate the first recourse of every professional who needs to know something – ask a colleague*
- *Setting up a new wiki knowledge base to make it easy for everyone to record new knowledge as it arises, and find it quickly when they need it.*

Practice profile (2005)

Staff: 20 partners, 95 staff
Offices: 2 (Bath and London)
Services: Architectural design in higher education, schools, housing, workplaces, the culture, sports and leisure and public and community sectors, urban design and masterplanning
Web: www.feildenclegg.com

Case study themes: hindsight review, networking directory and knowledge base

---

Starting points

It took some time to translate into action the management team's recognition that knowledge management had become a key issue for the practice.

In late 2003, when FCBa started to become actively involved in Spreading the Word, learning and knowledge-sharing depended on:

- Individual learning and informal person-to-person contact, with implicit encouragement from the flat staff structure and the culture of involvement
- Several special-purpose groups – Research and Innovation, Team Leaders, Design and Practice – charged (amongst other things) with disseminating knowledge within the practice. Most but not all of the group members were partners
- Annual practice awaydays for all staff, well away from office distractions in venues such as Barcelona and Amsterdam. These included 60-second presentations on all the past year's projects
- A basic knowledge portal, maintained by just one person
- An embryonic skills database.

Knowledge management was seen as largely synonymous with IT systems, but a review of available 'knowledge management' software had suggested that costs would be high, and led only to a decision to keep a watching brief.

Contact with Spreading the Word convinced FCBa that it needed to take a more holistic approach to knowledge management, that that is not only – or even largely – a matter of IT, that the software need not be expensive, and that all staff need to 'own' the system and be actively involved in developing as well as using it.

Hindsight reviews

FCBa decided that hindsight reviews offered the most promising way to learn more from its projects, and tried out the approach on one of its largest and most prestigious recent projects, the £30 million West-field Student Village at Queen Mary, University of London (QMUL). The largest student campus in London, this was designed to provide over 2000 bedrooms on a site bounded on one side by the Regent's Canal and on another by the main railway line to Liverpool Street station. FCBa became involved in 2001, construction started in late 2002, and the first two phases were occupied by the beginning of the 2004/5 academic year.

With the client's enthusiastic support, FCBa organised the first of two hindsight workshops immediately after the first two buildings had been handed over to Queen Mary in January 2004. Two more build-ings were close to completion, and work on a further two had started. The aims were to 'learn what has gone well and identify where we should improve for phase 3' and so 'to improve design quality through positive and creative feedback'.

The workshop was attended by seven FCBa staff – including the whole design team and senior partner Richard Feilden – together with seven other key people from the QMUL projects and accommodation management teams, the structural and building services engineers, and the quantity surveyors. The contractors were not represented because the contract was in process of being re-tendered for phase 3. The FCBa partner responsible for the project facilitated. Partici-pants were first asked to spend 20 minutes reflecting individually on their experience in the project and making notes on Post-Its of the lessons and problems that seemed most important to them. These were then discussed by the whole group for a further 2–3 hours.

After the workshop, the project architect reviewed the Post-Its and her own notes of the discussion, and distilled the lessons and issues onto two A3 pages under 13 headings: Brief/ERs, Fees/Scope of services, Quality and value, Cost, M&E, Procurement, Contractor/ programme, Quality control on site, Team, Landscape, Handover day,

The result – positive feedback, and The result – negative feedback. Finally, she extracted a short list of key actions for each party. The collated notes and key actions were sent to all participants.

Everyone was convinced that the event had been valuable, both personally and as a contribution to the success of phase 3. It gave all the participants an opportunity they would not otherwise have had to:

- Think through their own experience, and understand more clearly what lessons they had learned personally in the project so far
- Benefit from lessons learned by all the other members of the design and construction team
- Understand each other's perspectives and problems
- Raise potentially contentious issues in a safe environment where they could be discussed frankly and without risk of damaging working relationships
- Test ideas for improvement in debate, and gain confidence in them.

As a direct result, several aspects of the contract framework, programming and design for phase 3 were changed.

Queen Mary's project manager was so pleased with the workshop that he organised a second review in January 2005, after tendering for phase 3 was complete, so that the phase 1/2 contractor could be brought into the process.

Learning lessons from the first workshop, QMUL commissioned an independent facilitator to lead the event and leave the whole project team free to engage in the discussion. Post-Its were abandoned as too difficult to see, and were replaced by slides of key issues prepared by the facilitator from preliminary one-to-one conversations with the participants.

Twenty people took part, including most of the participants in the first workshop and key staff from the main contractor, the M&E contractor, and the building control consultancy. Even though it was held in the evening – and despite there having been very real strains at times during the project – discussion was friendly and constructive, and carried on for over three hours, beyond the planned finishing time. The whole event was recorded on minidisc, though in the event the recording was not used, and the lessons summary was based simply on the facilitator's notes.

Participants were as pleased with the experience as they had been with the first workshop: the QMUL project director was heard to say 'I'm completely sold on hindsight reviews – I'll be using them on all my big projects now.' Bringing the contractors into the review process introduced a new dimension, and led to new insights for many – particularly members of the design team.

Yellow Pages

When a dozen or two people work in one office it is possible for everyone to know everybody else and what their skills and experience are. Everyone knows who to ask when they need information or advice. But when organisations grow much beyond this size – and particularly when they grow fast, workloads are demanding and people are split between two or more offices, as at FCBa – it becomes increasingly difficult to know everyone, and the flow of mutual knowledge through ad hoc, person-to-person contact can dry to a trickle. FCBa was keen to keep it going, using IT to compensate for the inevitable decline in personal contact.

Like many organisations, its first thought was to develop a straightforward skills database. Participation in Spreading the Word persuaded it to think more deeply. Experience elsewhere with skills databases has often been discouraging. The information is usually too dry and stereotyped to be helpful, it can be difficult to relate to other sources such as project records, self-assessment of skills can be erratic, and entries are rarely kept up to date.

To avoid these problems, FCBa redesigned its embryonic database to:

- Be accessible through a web browser (instead of Microsoft Access) and at the hub of their intranet
- Link dynamically to their existing personnel, project and slide databases. At a stroke, this made it a much richer source of information, less reliant on people's personal input, and avoided the problem of keeping data in separate databases in synchrony
- Emphasise actual experience and willingness to help colleagues rather than potentially contentious judgements like 'expert' or 'novice'
- Give it a human face, both literally with a home page made up of staff portraits, each a clickable link (alongside a conventional drop-down list of names and a search box), and figuratively by encouraging people to include personal as well as professional information.

The system was also given a key role in personnel management, using it to target CPD and select 'topic champions' – people nominated as prime sources of technical advice – and plan to make it a key reference in annual reviews.

The result is a networking tool that works: user-friendly, information rich and adequately up to date. FCB has found that it helps to restore the informal knowledge exchange that used to be routine when the practice was smaller, and it has proved unexpectedly valuable for management as well.

Feilden Clegg Bradley Architects LLP

Figure 21.1 The home page of Feilden Clegg Bradley's *Yellow Pages* networking tool. The thumbnail portraits are all clickable links, and there are search boxes for names and projects.

Knowledge base

When practice manager Chris Askew first met the concept of wikis – websites built by their users – at a Spreading the Word workshop he seized on it as an ideal software framework for the practice's intranet. Three months later, FCBa's new system went live.

For FCBa, the key attraction of wiki software is its match to the practice's traditionally collaborative, inclusive approach to knowledge. Wikis provide all the functionality needed to make information quick and easy to find, and they allow anyone to contribute new material, at any time, using a standard web browser and simple syntax that can be learned in minutes. This is crucially different from conventional websites, which can be extended or updated only by staff skilled in HTML or specialist web design software. FCBa had found it impossible to keep a knowledge portal that depended on special skills and software up to date, and the need for new content to pass through one person distanced the system from users, and seriously inhibited contributions.

It was also important that wiki software be well proven, open source (and hence free), straightforward to set up, flexible, easy to integrate with other knowledge resources such as the practice's existing databases, and maintain a full audit trail of changes.

Once the decision had been taken to create a wiki-based intranet and knowledge base, and the broad architecture had been agreed, the foundations were laid very quickly. The basic wiki software was set up in a few days (by a junior architect with an interest in IT) using TWiki, one of the more widely used of several comparable wiki packages. The intranet home page, containing a skeleton framework for the main content of the knowledge base and links to the practice's pre-existing databases and other resources, followed in about another fortnight. Since then, fine details such as the design of the home page (first impressions matter!) have been gradually refined, and steady progress has been made to add knowledge content and link it extensively to additional internal and external resources. The software allows users to create web pages on new topics at will. These are automatically linked into the rest of the site, entered into the site map and full-text-indexed, but nevertheless unrestrained proliferation of topics can make a knowledge base unnecessarily difficult to use. To avoid this, FCBa planned a basic topic structure with care, initially with empty pages where no content was immediately available. To maintain control of the basic structure, write permission to the six top-level topic contents pages is restricted to the management team. Lower down in the topic hierarchy all users are free to add new sub-topic pages as they wish.

Users are all also free to amend or add to existing sub-topic pages, under the editorial control of the relevant topic champions. In the first year about 40% of staff made contributions – a typical rate for wiki knowledge bases.

After several months' evolution and refinement, the basic structure of the knowledge base settled down, with direct links on the home page to:

- The knowledge base proper, divided into six main topics: Buildings, Materials, Environment, Practice, IT Technical and FCBa Community, which contains an events calendar and news
- Four databases: Skills (the networking directory), Projects, Images and Certificates
- Administrative tools such as practice procedures and time sheets
- Key external websites, including Ribanet, Technical Indexes and FCBa's public site
- Site and web search engines.

The home page also carries the latest items of practice and personal news, and links to a 'sandbox' where new users can learn safely how to add material to the knowledge base, and to a range of technical management tools including a site change log and usage statistics.

The wiki software has proved to be trouble-free. Some human problems remain – interest in contributing or accessing knowledge is less widespread than originally hoped, and even the simplicity of wiki technology defeats some users – but FCBa is convinced that it was the right platform for its intranet and knowledge base.

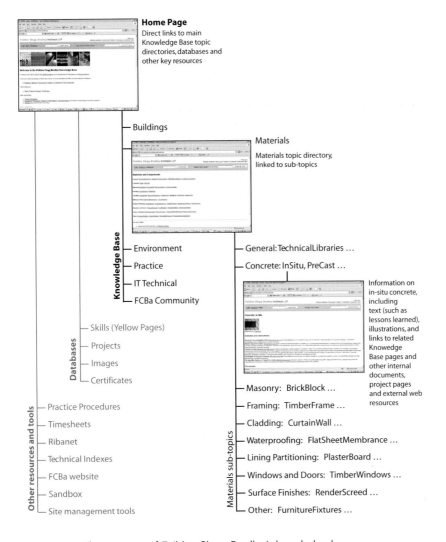

Home Page
Direct links to main
Knowledge Base topic
directories, databases and
other key resources

Buildings

Materials
Materials topic directory,
linked to sub-topics

Knowledge Base

Environment

Practice

IT Technical

FCBa Community

General: TechnicalLibraries ...

Concrete: InSitu, PreCast ...

Information on
in-situ concrete,
including
text (such as
lessons learned),
illustrations, and
links to related
Knowedge
Base pages and
other internal
documents,
project pages
and external web
resources

Databases

Skills (Yellow Pages)

Projects

Images

Certificates

Masonry: BrickBlock ...

Framing: TimberFrame ...

Cladding: CurtainWall ...

Other resources and tools

Practice Procedures

Timesheets

Ribanet

Technical Indexes

FCBa website

Sandbox

Site management tools

Materials sub-topics

Waterproofing: FlatSheetMembrance ...

Lining Partitioning: PlasterBoard ...

Windows and Doors: TimberWindows ...

Surface Finishes: RenderScreed ...

Other: FurnitureFixtures ...

Figure 21.2 The structure of Feilden Clegg Bradley's knowledge base.

Commentary

FCBa's experience with hindsight reviews and their networking directory and knowledge base contains numerous lessons applicable in any practice.

It confirms the experience of other kinds of organisation (summarised in Chapter 25) that hindsight reviews are time well spent when they look at suitable projects, and are well planned and well conducted. Participants enjoy and value them. It can be difficult to bring people together from several organisations, but doing so adds considerable value. Juxtaposing the experience of the different parties can reveal important lessons that personal and single-organisation reviews would miss, and as a bonus the experience often improves inter-organisational relationships as well. Twenty participants are too many, particularly with seating around a long table, which makes it difficult for everyone to see everyone else. With so many, conversation becomes noticeably more formal.

To get the best value from them, hindsight reviews need to deal with suitable cases for review, and the participant list has to be appropriate. Circumstances that tend to increase the value of a review include:

- Profitability, client satisfaction or other key success measures either markedly above or below expectations
- A significant aspect of design, design method or project organisation that is relatively new to the practice
- Significant difficulties
- High risk
- Likelihood of significant further work of any kind for the same client
- Likelihood of significant similar work for other clients.

The first four of these make it likely that there are significant lessons to be learned, and the last two that learning them will have real business value. The Queen Mary project was an ideal case because, as it was a very large and complex project (by FCBa standards) divided into three parts, there was certainty that lessons learned after the earlier parts would have immediate and important application. Relationships had also been strained at times – a significant difficulty. It was not easy, though, to decide who to invite. There were so many parties involved that including representatives of all of them at various levels (as is desirable, to ensure that first-hand accounts are available of all the issues that come up in discussion), including the whole FCBa team, threatened to make the numbers unmanageable. In the event the absence of the contractor contingent kept the first workshop within bounds, but numbers grew to 20 in the second, and the quality of discussion suffered accordingly.

At a detailed level, the Queen Mary experience confirms that it helps to have an independent facilitator, and to structure discussion around a timeline and key issues identified beforehand. As well as unburdening a key participant, this makes for a more considered list of issues, and thinking about them beforehand makes the participants better prepared for fruitful discussion.

FCBa's approach to the design of its networking directory and knowledge base was exemplary. Notable touches include:

- The use of thumbnail staff photos on the home page of the networking directory, both as a humanising element and as one of the several means of accessing individual entries
- The stress on actual experience and willingness to help, and avoidance of difficult judgements such as 'expert' and 'novice'. These are a common feature of skill-finder tools, but they are fraught with pitfalls. It is not easy to define objective criteria for distinguishing between skill levels; self-assessment is unreliable – the overconfident overrate them-selves, the diffident underrate – and a poor indicator of ability or inclination to be helpful; and assessment by others can lead to resentment. In practice, most of the questions people want to ask are mundane, and practical experience and willingness to be helpful are more valuable qualities than nominally high levels of expertise.
- Using the networking directory as a principal source of information in personnel management. This gives staff an incentive to update their entries (if only once a year, before their annual review), and gives line managers an interest in maintaining the system.
- The imposition of a carefully considered basic structure in the knowledge base, editable only by the management team. This helps to avoid the descent into chaos that let to the abandonment of ECA's wiki.
- Minimal barriers to contributions at lower levels in the hierarchy.

A final point of interest in FCBa's experience is that the rapid devel-opment and thoughtful design of the IT tools are due largely to its having a practice manager with the imagination, commitment and authority to make things happen, and with the deep knowledge of the practice that is necessary to make it successful. It helped, too, that the practice had a member of staff with the technical ability to imple-ment the wiki, and a willingness to try new things. The success of knowledge management initiatives often depends more on the ability and drive of one or two key people than on anything else.

Chapter Twenty-Two
Case Study: Penoyre & Prasad

After more than a decade of slow and steady growth since its foundation in 1988, Penoyre & Prasad has burgeoned in the past few years:
a team of 35 in 2001 almost doubled by 2005. The partners – by then
nine instead of two – were determined to retain the practice's ethos
and its approach to design, but they had to rethink the management
systems radically.

The rethink included the practice's attitude to knowledge. It was
clear that staff could no longer rely on osmosis to learn how to do
things, and traditional knowledge resources – a library, subscriptions
to information services such as Barbour Index, a hard-copy 'book of
details' and a variety of largely unconnected databases and electronic
document files – were no longer enough. The practice plan of March
2003 said: 'It is now widely understood that in any organisation the
quality of flow and transfer of knowledge in its many forms is a key
element of high performance. Staff, associates and partners in the
practice have identified a weakness in this area. We know there is a
lot of knowledge locked up in individuals which if more widely shared
could dramatically improve the practice's capability. The growth and
management of knowledge in the office is our best counter to the
threat from our competitors.'

Practice profile (2005)

Staff: 62
Office: 1
Services: Architectural design, primarily in the health, education and
            arts sectors
Web: www.penoyre-prasad.net

Case study theme: knowledge base

Starting points

By the time Spreading the Word started, P&P had already taken several new knowledge initiatives. These included regular lunchtime CPD sessions on Friday lunchtimes – 'sometimes wonderful, often satisfactory and very occasionally awful' – and 'management groups' of enthusiasts, each headed by an associate, with briefs to take the lead in specific areas such as R&D, legal and professional issues, CPD, IT, and marketing. With no time budgets the groups met only irregularly, but they gradually worked through their self-generated task lists and got the systems working better.

The R&D group's work quickly began to centre on knowledge management (indeed, it was later renamed the knowledge management group), and it took two further initiatives: focusing the Friday lunchtime sessions more clearly on learning and knowledge-sharing, and creating a simple 'R&D database' – the practice's first attempt at a knowledge bank.

Some of the Friday sessions were used to analyse the major elements of a building, with P&P's own completed buildings as examples. These generated a lot of interest. Staff enjoyed them and found them useful, and the written-up notes were expected to become a valuable resource. In the event, though, the sessions stopped after one tour through the principal building elements, and the notes languished in a file, largely unread.

The R&D database disappointed, too. This was intended to mimic the learning that used to happen naturally when experienced staff overheard telephone conversations or chats by the photocopier, and offered advice. The thinking behind it was that if interesting nuggets of knowledge gained from project work or research could be captured, the database could become the first port of call for advice, and staff would no longer need to jump up and ask everyone within earshot how to detail a roof membrane abutment. But that did not happen. Technically the database worked well, but despite two presentations to the whole office and a publicity drive only 35 entries were made in the first 10 months, most by the database's developer, and far too few to attract significant use.

The experience with the new-format CPD sessions and the R&D database – together with input from Spreading the Word – convinced the R&D group that there was little value in creating small, isolated caches of codified knowledge. Knowledge needed to be brought together in one resource where it would be readily accessible and have the critical mass needed to make it self-evidently useful, and keep it in constant use. Thinking about knowledge management needed to become much more joined-up.

The R&D database

It was not clear what was wrong with the R&D database. The developer, an enthusiastic young architect, had little experience of IT and none of knowledge management, but he had nevertheless set about the task sensibly. With the R&D group, he formulated a series of basic design principles. The database should:

- Be designed principally for recording technical knowledge that is easily summarised
- Treat knowledge as the distillation of a conversation
- Record knowledge in distinct, concise snippets, each written and owned by a single author
- Be simple and stand-alone, operating independently of other office systems.

A review of software options led to the choice of FileMaker Pro software, and a single, simple form was developed for entering, viewing and printing records. This included fields for:

- The CI/SfB[1] code and label
- A title and subtitle
- A 'commentary' of up to 300 words – the main content
- 'Supporting information' including material, product, manufacturer, supplier, contractor, project reference, project name, author's name and date
- Up to three contacts and three document references.

The failure of the database led to some soul-searching: the balance between effort and reward was clearly unattractive to potential users, but why? To gain some insight into this, the entire staff were surveyed by questionnaire. About half responded, and the feedback was forthright and revealing. Staff felt that:

- Conversation was still more useful than electronic records.
- The format was dull, and too geared to technical knowledge. The system also needed to accommodate other material such as images in order to appeal to architects.
- The status of records was unclear – were they simply individual experience or office practice?
- Content needed to be more selective.
- The system needed more support from senior staff.
- It was too easily forgotten: people needed constant reminders that it existed.
- There should be a time budget for recording knowledge.
- People should be asked to contribute to specific topics.

---

[1] The international standard classification system for construction products.

- The database structure was unclear.
- Access needed to be faster and easier.

Some of the reactions were consequential rather than causal – the database would not have been forgettable if it had been really useful, for example – but overall the results helped show what needed to change. At about the same time, a Spreading the Word workshop gave P&P an opportunity to compare notes with other practices that had developed knowledge bases (or had tried to). After considering all this new information, the R&D group decided to stay with the familiar technology of FileMaker Pro, but to use it in a more sophisticated way to create a more flexible, user-friendly and attractive system, and back it with more resources and sustained encouragement to users.

The knowledge bank

In two months, the database was completely redesigned. Thanks to the use of simple, familiar software, the cost was modest: Penoyre & Prasad estimates that developing both the original R&D database and the knowledge bank took only about 19 person-days. The new system could accommodate a spectrum of design knowledge, and link to images, drawings, documents, videos, and external websites. Its potential was clear – but it still needed content.

One of the key lessons from the R&D database was that contributions are not naturally forthcoming – people are variously too busy to contribute, too shy, feel they have nothing to offer, or simply lack clear motivation. The R&D group decided that content must be explicitly elicited, at least until it reached critical mass. That is a common experience with knowledge bases. People will put knowledge in only if they feel they are being repaid by the knowledge they get out, or if they get some other psychological reward such as recognition. To encourage contributions and a sense of ownership of the system, Penoyre & Prasad conducted a programme of one-to-one interviews. People were asked to make at least one contribution on:

- An aspect of design, construction or practice, such as natural ventilation in schools or stabilised soil blocks.
- A finding from research in the office, for example on non-slip floor finishes in health care facilities
- A finding from experience on site, such as how to achieve a good finish to fair-faced concrete, or
- A reference to a particularly useful external source of information, knowledge or guidance.

It was stressed that contributors did not need to be experts, but should simply have some knowledge that colleagues were likely to find useful.

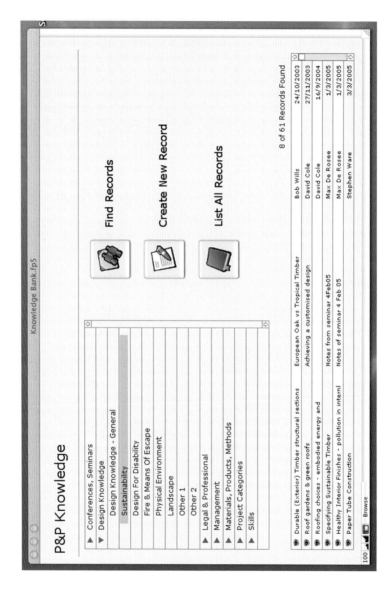

Figure 22.1 The entry screen in Penoyre & Prasad's knowledge bank.
The top part of the screen contains a hierarchical topic directory, and buttons to open a search dialogue, a data entry form, and show all records. A scrollable hit list of records on a selected topic (here 'Design Knowledge – Sustainability'), found by a word search, or in the whole database appears below. Items in the list have clickable links to the individual records.

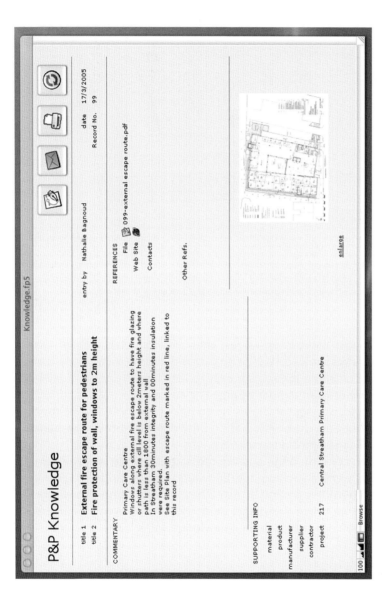

Figure 22.2 A typical record.

Individual records include most of the same fields as the original R&D database: a title and subtitle, contributor name, date, commentary (the main content), and supporting information including material, product, manufacturer, supplier, contractor and project. New additions include fields for links to separate files and to websites, contacts and other references, and illustrations.

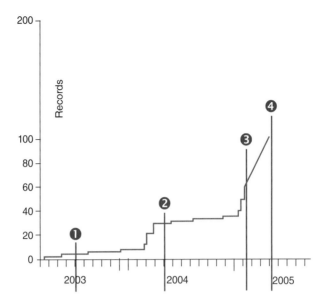

Figure 22.3 The new knowledge bank has attracted many more contributions than its predecessor.
The original R&D database was launched with just four records, and grew to only 30 in its first year. A year later its contents – now 58 records – were transferred into the knowledge bank, and users added another 43 in the following three months.

A second user survey, six weeks after the formal launch of the knowledge bank, showed that people found the new structure and interface much clearer, the system easy to use, and the content interesting. They were keen to hear by email when new items were added, and they would like sector teams to contribute knowledge of their particular fields, and closer integration between the knowledge bank and other office systems and culture.

Observation confirmed the positive message from the survey. Many more people were searching the knowledge bank and contacting contributors to ask for further advice. They were quoting knowledge bank material at office meetings. And they were even overheard saying 'You should put that in the knowledge bank'!

Lessons learned

Penoyre & Prasad drew several lessons from their journey from the R&D database to the knowledge bank:

- Terminology is important: it found that people respond better, for example, to the idea of knowledge-sharing than to knowledge *management*.

- A knowledge base needs a designated 'editor', and his (or her) role is vital.
- At the same time, people must be convinced that everyone is free to contribute.
- Visible links between knowledge sources and activities – knowledge bases, office meetings, seminars and so on – are vital.

Knowledge management is not a simple matter of buying software or building a database; knowledge initiatives need strong support from the top, serious and informed thinking, close attention to human factors, persistence, and, from time to time, critical reassessment.

The experience convinced P&P that it should continue backing and developing its knowledge bank, and start other knowledge initiatives to complement it. Amongst other things, it planned to start a programme of hindsight reviews, digitise its 'book of details', add a skills register of everyone in the office to the knowledge bank, and keep all the practice's knowledge-sharing activities in people's minds with monthly bulletins.

Commentary

Penoyre & Prasad's conscious awareness of knowledge, its importance, and the need to manage it started in the best possible way: from a major rethink about its future, and about its entire approach to the management of the practice. This convinced the partners that knowledge was central to realising their ambitions for the practice, and that they needed to manage it as consciously and systematically as other aspects of the business.

As they later found, though, awareness of a principle does not lead automatically to practical success. That requires a much richer and more detailed understanding of the issues, practical experience, and the determination to persist and try again when ideas fail to work out.

It is interesting that Penoyre & Prasad chose to use a conventional database for their knowledge bank rather than adopt a more modern platform such as a wiki. Edward Cullinan Architects, a practice of broadly comparable size and technical resources, has recently reverted to databases after a two-year trial with a wiki. Anecdotal evidence suggests that a preference for familiar database technology is not unusual in small practices, despite its technical limitations. The full reasons are not clear, but two factors appear to play a part:

- The preferences of the people responsible for implementing the system
- The greater thought involved in designing a structure for a platform as flexible as a wiki.

Users who have no interest in IT (including most architects) are easily baulked by small difficulties they encounter with unfamiliar software, and they are unforgiving of bugs and system failures. IT support staff quickly learn that they receive little praise when systems work, but plenty of brickbats when they don't. This biases both sides towards familiar software, even when technically superior but unfamiliar alternatives are available.

Once the basic platform has been chosen, further choices have to be made about how content is to be organised and accessed. These are generally easier with a database than a wiki because there are fewer options available, and once made they are more difficult to change. In principle, and in practice when enough technical expertise and management effort are available, flexibility is a strength, but when they are not it makes the structure vulnerable to misuse and can lead to chaos, as ECA found. Fortunately, people are remarkably accepting of constraints over which they believe they have no control.

Despite its limitations as the framework for a knowledge base, then, a simple database can be a rational choice for a small practice at present. As Barry Schwartz famously pointed out in *The Paradox of Choice: Why more is less*, satisficers (who are content with the adequate) are generally happier than maximisers (who invest heavily in effort and emotional energy to get the best).

In the future, wikis and related software promise to become a normal part of IT staff's expertise, and familiarity will cease to favour databases. The emergence of commercial wiki-hosting services, which avoid the need for in-house IT skills, is already removing one of the obstacles to their adoption. Usage of Wikipedia and other wiki-based systems continues to grow apace, and in due course it seems inevitable that end-users will come to see them as the norm, and become less tolerant of the limitations of databases.

Chapter Twenty-Three
Case Study: Whitbybird

Founded in 1984 as a specialist structural design consultancy, Whitby-bird has grown and diversified over the years into a highly regarded multidisciplinary engineering practice of 300 people. In 2004 alone it won 30 awards from the RIBA (including London Building of the Year), the RICS, the Royal Fine Arts Commission, the Civic Trust, the British Council for Offices, and others.

Over the years the practice has evolved a variety of procedures and IT tools to promote learning from experience and knowledge-sharing. By 2003 end-of-project reviews were well established and widely valued; the IT tools included a knowledge bank, staff skills and Who's Who? databases, and an email-based 'team briefing' system through which anyone at team leader level or above could broadcast new information to colleagues. Despite some failures – the knowledge bank, for example, 'fell apart' because it was too difficult to use – management regarded knowledge-sharing as generally effective, with good buy-in from staff.

But director Charles McBeath, who is responsible for coordinating the practice's IT systems, was keen to improve further. He could still see wheels being reinvented, and he wanted to bring the engineering, management and administrative aspects of the practice's work closer together. He believes that 'The use of knowledge defines a company's culture – the management of knowledge defines its success.'

Practice profile (2005)

Staff: 300
Offices: 6 UK, 1 Dubai
Services: Engineering consultancy: structural, building services, fire,
            facade, geotechnical and infrastructural and urban
            engineering, bridge design, special projects, community
            energy, and sustainability and renewable energy
Web: www.whitbybird.com

Case study theme: knowledge audit

As a first step towards achieving Charles McBeath's vision, Whitbybird decided to carry out a simple knowledge audit to identify some of the strengths and weaknesses of the existing systems, and provide some starting points for improving them. The audit was based on a questionnaire survey of 100 of the younger engineers, and focused on their perceptions of the four knowledge systems that the Operational Process Management (OPM) team and the board judged to be the most important:

- Company communication strategy
- Task groups (Whitbybird's name for communities of practice)
- The Who's Who? database
- The online feedback system.

The response rate was good – 77% – and the OPM team, who were given responsibility for overseeing the audit, were satisfied that the results usefully clarified understanding of these systems, and pointed the way to worthwhile improvements.

The process involved seven main steps:

1 Identifying the systems and assets that contribute to knowledge management
2 Selecting a subset for audit
3 Choosing audit methods
4 Designing the questionnaire
5 Testing and refining the questionnaire
6 Conducting the survey
7 Analysing the results.

1: Identifying knowledge systems and assets

Between them, the board and the OPM team identified 11 principal knowledge systems:

- Company communication strategy
- Design critiques and technical reviews
- Online quality management system
- Task groups
- Online feedback system
- Technical highlights
- Training/CPD/PDR
- Who's Who
- Staff induction and mentoring
- Job management system
- Companies and contacts database

and five knowledge assets:

- Staff
- Work processes

- IT network
- Databases
- Library.

2: Selecting a subset for audit

Investigating 16 systems and assets would have required more effort than the board wanted to commit to the audit, so they decided to limit its scope. Subjective assessments of the importance of the various systems to the company and their potential for improvement led to four of them being selected for investigation: the company communication strategy, task groups, the Who's Who, and the online feedback system. The board considered the communication strategy (an unusual target for a knowledge audit) to be a key system, because they saw it as the main influence on how well staff understood Whitbybird's ethos, organisational structure and knowledge systems, and how well corporate information and management decisions were communicated to staff. The task groups – broadly equivalent to communities of practice – are the practice's chief mechanism for responding to feedback, recommending changes in operational and technical procedures, and highlighting issues that need further attention. The Who's Who system plays a key role in the sharing of tacit knowledge by helping staff to discover who knows what, and how to get in touch with them. The online feedback system is an important mechanism for bringing personal lessons learned on business processes into the knowledge system. Contributions are monitored by a feedback review group, who decide what action should follow.

3: Choosing audit methods

When knowledge systems disappoint, it is often because their design fails to take account of the realities of the corporate culture and of people's working patterns and motivations. Having decided to devote only a small amount of time to the audit, Whitbybird chose to base it on a survey of people's awareness and perceptions in order to get the quickest possible insights; objective measures such as who uses systems and how often (by analysis of server logs, for example) were rejected on the grounds that they expected them to yield relatively little with the effort available. Resource constraints dictated the survey method, too. Interviews were ruled out because only a few could have been carried out and analysed in the time available, whereas the practice already had an online questionnaire system that would allow a substantial proportion of staff to be surveyed in a short time, and would calculate results automatically.

Recognising that a survey can itself help raise awareness of the issues it covers, the OPM team decided to concentrate on the younger engineers and send the questionnaire to 100 of them, evenly sampled from the different engineering teams.

4: Designing the questionnaire

Questionnaire design has to balance richness and detail of enquiry (which requires many questions) against response rate (which suffers if there are too many). Whitbybird decided to favour response rate, aiming to make the questionnaire short enough to be completed in 5 minutes. To collect as much information as possible with this very tight constraint, the questionnaire included a mixture of 'closed' questions (questions that require respondents to select from a list of pre-defined answers) and 'open' questions (which invite respondents to write in whatever they like).

The questionnaire focused on three issues: how much each of the four selected systems (the company communication strategy, task groups, the Who's Who and the online feedback system) was used, how effective it was perceived to be, and how it could be improved. It included between three and five closed questions about each system to probe use and effectiveness, and a single open question inviting suggestions for improvement. Most of the closed questions had four or five response options based on logical alternatives (such as 'none', '1', '2', '3' and '4 or more' for 'How many task groups are you currently a member of?') or on subjective scales (such as 'difficult', 'could be improved', 'OK' and 'very easy' for 'How easy is it to find relevant feedback items?').

To help respondents understand the questions (and at the same time educate them about the knowledge systems) each group was preceded by a short description of the system and its purpose. The whole questionnaire was implemented as a series of online forms feeding responses into a database, ready for analysis.

5: Testing and refining the questionnaire

Even with skilled design, questions can still fail to elicit useful responses: it is easy, for example, to miss an ambiguity that causes some (or all) respondents to answer quite a different question from that intended. Pilot testing is invaluable for weeding out problems such as this, and it can also provide a useful check on completion times and the mechanics of survey forms and data-handling code. Accordingly, Whitbybird sent the questionnaire first to ten respondents variously selected for their known eye for detail, diverse opinions, and just at random to gauge uninformed reaction. Some questions were modified in the light of the returns.

6: Conducting the survey

The final survey was sent to 100 young engineers, with a two-week deadline to respond. Nearly 80% did so.

7: Analysing the results

The responses to the closed questions were analysed by proportion and plotted as pie charts, and the free text responses were reviewed by the OPM team, and recurring themes and comments were identified.

The results were broadly consistent with the team's expectations, showing that all four systems were well used and generally effective, but that there was clear scope for improving them. The team were satisfied that the survey as a whole gave valuable support for the case for further development, and the responses to the open questions gave useful pointers to specific problems and ideas for specific improvements. Overall, Whitbybird judged that the value of the results more than justified the audit, and the experience gave them a valuable insight into how an audit could be used to help improve other aspects of their knowledge management in the future.

Two specimen results are shown in Figures 23.1 and 23.2.

Sample results from Whitbybird's knowledge audit

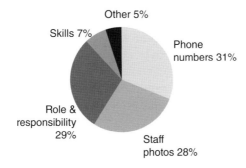

Figure 23.1 Asked 'What do you most use the *Who's Who* for?' in Whitbybird's knowledge audit, 31% of staff said they used it most to look up phone numbers, 29% to look up people's roles and areas of responsibility, 28% to look at staff photos, 7% to look for people with specific skills, and 5% for other purposes.

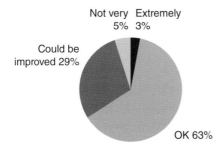

Figure 23.2 Asked 'How effective do you think task groups are at gathering knowledge and driving continuous improvement?', 63% of staff said they were OK, 29% that they could be improved, 5% not very and 3% extremely.

Commentary

Whitbybird's knowledge audit was heavily constrained by resources and the decisions to use only a questionnaire, and to minimise the demands on staff time. As a result, it was limited in what it could achieve.

With the opportunity to ask relatively few questions, the preponderance of those asking for broad, qualitative judgements ('How effective is . . .') is understandable. However, questions like these are not diagnostic. This does not matter when the consensus opinion is strongly favourable, but it does in the more usual case when opinion is mixed or clearly *unfavourable*. Broad quality judgements give no indication of causes, and so no indication of how systems and procedures could be improved. In addition, the significance of results is clouded by uncertainty about how respondents have interpreted 'effective'. Even short surveys can be made more informative by focusing on key aspects of performance, and on more objective measures such as frequency and purpose of use. With careful framing, questions can give useful insights into specific aspects of effectiveness, and ways in which it could be improved.

The two specimen results from the Whitbybird audit illustrate the difference between these two approaches. The first, which asks respondents about the effectiveness of task groups, produced the uninformative, middle-of-the-road response that questions like this typically elicit. A similar question about the Who's Who might well have produced a similar result, but, as the second pie chart shows, asking respondents what they use it for is much more revealing. The responses show that it is used most often as an internal phone book (and is probably effective in that role), but relatively rarely for its intended purpose of locating people with specific skills, possibly (but not necessarily) because it is not effective for that. Follow-up questioning could then investigate whether this is because people only rarely want to locate skills or because the system fails to give helpful results, and if the latter how it could be improved.

When resources allow, more extensive audits involving interviews and a longer questionnaire normally offer better value. A response rate of over 40% can usually be achieved even with 50–60 questions, provided they are carefully framed to avoid irritating respondents. This is enough to ask diagnostic questions about the effectiveness of general processes (such as learning from experience and sharing knowledge with colleagues) and of several specific systems and procedures, and to cross-check the results with questions about 'litmus test' cases. More detailed investigation like this is also more revealing of differences between offices, grades of staff and other significant groups.

Chapter Twenty-Four
Case Study: WSP

WSP is a good example of a company that has a great deal to gain from knowledge management, is aware of this at board level, and yet – on its own admission – is finding it difficult to become the kind of learning organisation it would like to be.

The potential benefits of good knowledge management in a multi-disciplinary consultancy as large and geographically dispersed as the WSP Group are obvious. Even the 200 or so major projects they have in hand at any one time in this country – to say nothing of many more overseas – provide rich opportunities to learn lessons that have the potential to lead to better solutions and more efficient working in the future.

Practice profile (2005)

Staff:	1850 UK (5300 worldwide)
Offices:	24 UK (100+ worldwide)
Services:	Development planning, building and infrastructure design, environmental and geotechnical studies and network and facilities management for the property, transport and utility sectors
Web:	www.wspgroup.com

Case study theme: technical coordinator workshops

Starting points

Knowledge management strategy and implementation in WSP is left to the discretion of individual businesses. In the UK it is most fully developed in the property sector business, where it is led by two full-time staff in the Group Technical Centre (GTC) in London. The GTC has built up a range of procedures and IT tools and an extensive library of electronic reference material, and is supported by two engineers in each office, who act as links to local staff. The brief of these local technical coordinators – one building services and one structural engineer in each office – is to act as points of contact between their offices and the GTC, feed material from the GTC out to local staff, feed back local lessons learned, promote use of the knowledge base, and suggest improvements to the system.

By 2005 the main elements in the knowledge system were:

- A technical reference manual library of around 250 documents, together with bought-in libraries, specifications etc., accessible to all staff over the intranet
- A monthly group technical bulletin produced by the GTC, which includes 'watch-it' notes alerting staff to potential pitfalls
- A roadshow of seminars led by the GTC, which visits each office every six months or so
- Occasional one-day get-togethers for specific professional groups
- A technical 'help desk' service provided by the GTC
- A multi-topic discussion forum that automatically emails members when new items appear
- An electronic telephone directory (to which photos were recently added) and a skills register, which staff are required to update annually.

With this infrastructure in place, Group Technical Coordinator Stuart Alexander turned his attention to increasing buy-in and usage at the grass roots. In a big and busy company like WSP well-proven working methods have considerable inertia, and it takes time and persistence to make systematic learning and knowledge-sharing a habit for everyone. As recently as 2003 local teams and offices still tended to develop their own solutions, with little reference to experience elsewhere. It was hard to find staff who were willing and well qualified to be local technical coordinators, and whereas experience in other industries suggests that the most valuable knowledge-sharing often occurs in casual conversation with colleagues, people found this difficult in the company's head-down, time-conscious culture. All this was changing slowly, but even in 2005 the value of learning from experience was still not fully recognised at local level, and knowledge flow remained largely one way – from the centre outwards. The GTC estimated that the proportion of staff who were active knowledge contributors increased from about 3% to 15% in the two years from

2003 to 2005, but it wanted to increase this further, and destroy the silo mentality for good.

Technical coordinator workshops

The Group Technical Coordinator's first major initiative to encourage more learning and knowledge-sharing at a local level was to develop a 'project life cycle' procedure that includes provision for foresight and hindsight reviews. This became mandatory in 2004. In parallel with this, he initiated a series of regional technical coordinator workshops to show the office coordinators how valuable local learning and person-to-person knowledge networking can be.

The workshops aim to foster a sense of community among the local coordinators and give them an opportunity to:

- Compare experiences and discuss concerns with each other and with the GTC team
- Learn about developments in the knowledge infrastructure
- Learn about specific 'hot topics' in their discipline so that they can act as local experts.

From the GTC's point of view, the workshops are also an opportunity to get feedback on other components of the company's knowledge infrastructure. To broaden involvement, each local coordinator is invited to bring a second person from their discipline, and centre heads are also invited; the GTC team provide the chairman and secretary. Four of these workshops – all half-day events starting with a sandwich lunch – were held in the first year, each at a different location and covering three or four offices. The response was encouraging. The local coordinators welcomed the opportunity to get together, and they scored the events highly on feedback forms, with the overwhelming majority of assessments 'good'. Technical coordinator workshops have since become a regular part of the GTC's programme.

The introduction of the technical coordinator workshops and the project life cycle procedure – which will make it harder for busy engineers to treat knowledge activities as a disposable luxury – marks an important stage in the evolution of WSP's knowledge management system. They give people at the grass roots opportunities to see for themselves what a systematic approach to learning from experience and knowledge-sharing involves, and the personal and corporate rewards it can bring. And that, in turn, should help extract increasing value from the other parts of the company's knowledge infrastructure. Nevertheless, the GTC recognises that without other and more radical changes it may be many years before the culture becomes truly knowledge conscious, and before self-sustaining, user-driven knowledge systems can be established successfully.

Commentary

Size makes the business benefits of good knowledge management clearer, and dedicated staff affordable, but, as WSP's experience shows, these are not unmixed blessings. Full-time knowledge managers are rarely appointed at partner or board level, and as a result they lack the authority to commit other resources, take quick decisions or drive through obstacles. This drains credibility from assertions that knowledge is strategically important to the organisation: actions speak louder than words. It is also a false economy. Unless the knowledge staff are supported by a long-established knowledge culture (as they are at Arup), progress is apt to be slower than in smaller and more agile practices, and the opportunity costs can be large.

The Group Technical Coordinator's initiative appears modest, but it was well judged. Whether or not there is strong and credible support from the top, progress always depends on the knowledge and enthusiasm of local champions. Bringing them together for periodic workshops is an effective way to develop their knowledge management skills, encourage their efforts, and so stimulate learning and knowledge-sharing at the grass roots.

Chapter Twenty-Five
Case Studies on Foresight and Hindsight

Project-based working creates both difficulties and opportunities for organisational learning. These are magnified when there are several firms involved, and especially when the contractual framework is inimical to mutual trust (as it often is in UK construction). A strong focus on delivery and on time pressure (real or perceived) makes it almost impossible to organise specifically learning-focused activities while construction projects are under way. The reason most commonly offered in professional practices for limited engagement in knowledge activities of all kinds is simply 'I haven't got the time.' However, sentiment is more amenable in two short periods: the earliest stages of projects, and immediately after they have finished. People are at their most open-minded at the beginning of projects before their thinking starts to gel and limit the options they are prepared to consider, and this makes it the ideal time to introduce ideas from outside the project team and lessons learned in the past. Immediately after a project has finished, when the pressure is off, the next one has barely started, and memories are fresh, is an ideal time to reflect and learn lessons. As with other knowledge activities, a systematic approach to these two opportunities – foresight and hindsight respectively – pays dividends, and the case studies in this chapter show how a variety of different firms have used them to good effect.

The idea that useful lessons can be learned from post-completion project reviews is widely accepted, and in many professional practices they are required by QA procedures. However, anecdotal evidence suggests that they rarely happen in practice, and when they do they are often little more than ritual form-filling – an ineffective process leading to disillusionment and inaction in a spiral of decline. Hindsight is a very different process, as the summary in Table 25.1 shows.

Within this general framework (as we saw in Chapter 10) hindsight can take several forms, and the case studies illustrate several of these.

Typical post-project review	Hindsight
One firm	Several firms
Carried out by one person	Involves several people
Management only	Several levels
No protected time	Protected time
Reflection	Reflection and dialogue
Principally high-level assessment	Search for detail and causes
Brief, form-based record	Rich descriptive/analytic record
No knowledge resource produced	Knowledge resource produced
Only one person learns lessons	All participants and potentially all staff learn lessons

Table 25.1 Hindsight: very different from post-project review.

They also make the point that all the main design consultants and the client are equally well placed to initiate the activity.

Only one of the cases discussed in this chapter looks at foresight. This is still (even more than hindsight) an undeservedly rare activity in construction. This is surprising at a time when increasingly demanding requirements for cost control, risk reduction and environmental performance call for a more imaginative and multidisciplinary approach to design, and when new contractual frameworks are removing some of the barriers. The example here shows how powerful a technique it can be.

The cases in this chapter are:

- A post-project review workshop – hindsight – carried out by housing association Amicus to learn lessons from a development that had been dogged by relationship problems
- A combination of a workshop and interviews used by airport operator BAA to review the development of a new check-in area in an existing terminal building, and learn lessons about managing time-critical projects and carrying out construction work in areas used by the public
- BP/Bovis Global Alliance's use of a pre-project review exercise – foresight – to find radically new ways to build petrol stations and slash construction costs
- Engineering consultants Buro Happold's interview-based hindsight review of a museum project, in which the learning history approach was used to analyse the results and disseminate the lessons learned
- Contaminated land development specialists Lattice Property's use of a hindsight workshop to learn lessons from a

project where an unfamiliar, but potentially widely useful, process had been used for on-site remediation of polluted ground.

The conclusions at the end of each of the four hindsight case studies are the organisations' own.

Amicus Group

In 2003 housing association Amicus Group (now the AmicusHorizon Group) was managing over 15 000 homes in London, Kent, and Sussex, and typically building 100–200 more a year. It was a leading member of the Amphion Consortium, a 'club' of 17 housing associations set up to drive down construction costs by developing efficient prefabrication techniques and establishing a partnering relationship with a major developer, with members sharing the initial costs and pooling buying power. This enabled more strategic thinking than would have been possible under either project-specific or multi-project partnering arrangements at a local level.

Amicus first used hindsight to learn lessons from a small consortium development in Canterbury, Kent, where it had used a proprietary timber frame construction system that it was considering adopting on a wide scale. This project suffered from a succession of managerial complications. Amphion originally contracted with Beazer Partnerships in 1999 to supply the timber frames; Beazer was acquired by Persimmon Homes a year later; and the timber frame manufacturing business (Torwood Homes) was subsequently sold to its management and became Partnerships First in October 2001. The negotiations between Persimmon and Partnerships First, and the setting up of the frame factory, formed a backdrop to much of the project. This had an adverse effect on the resources available, on staff morale (Persimmon was downsizing parts of the Beazer organisation), and on the efficiency of supply routes.

The construction method was not the only innovative aspect of the project. With help from project management consultants Davis, Langdon & Everest and legal advisers, Amphion developed approaches that were new to them for pre-contract project management (including fast-track appraisal and open-book pricing) and for the construction contract itself. The partnering ethos allowed issues of timing, liaison with utilities and so on to be approached with a common purpose, rather than simply as problems for the contractor to resolve. However, when programmes slipped, the familiar problems of a blame culture – reluctance at site level to acknowledge delays, and consequential arguments about the timetable and costs – all resurfaced. The project finally completed at the upper end of the expected range of costs, and late.

The innovative aspects and difficulties of the Canterbury project, together with the fact that it was typical of the projects likely to follow in Amphion's development programme, made it an ideal opportunity for its first exercise in hindsight. It undertook a review based on:

- A desk review of project records
- A combined review and dissemination workshop for seven senior team members from Amicus, Partnerships First, Davis, Langdon & Everest and the legal advisers
- Individual interviews with people who were unable to attend the workshop.

The review was held one week after practical completion, while memories were fresh, without time pressure, and well away from office interruptions. All the invitees were asked to review their project records before the event to identify the events and issues that would most repay discussion, and a consolidated list was displayed on a flip chart to help guide the discussion, with space to note lessons learned. The workshop leader took care to pace the discussion, and to encourage everyone to contribute and to challenge and debate, starting with what happened, and then moving on to why it happened and finally to doing better. Despite the strains during the project, participants spoke remarkably freely and constructively. Numerous new insights emerged into why things had happened and how difficulties could be avoided in future projects.

The follow-up interviews each lasted about an hour, giving time for each interviewee's point of view to be fully explored and issues to be followed through. Amicus found that this provided a richer picture in some respects than the workshop discussion, but the issues that emerged were largely similar.

How it worked

Everyone involved found the structured approach helpful. The hindsight process teased out numerous lessons that would otherwise have been missed, and the effort paid dividends in new insights at all levels, from site to director, both about project specific issues and – perhaps more valuably – about how teams operate, particularly in a partnered environment. It also had a valuable effect on relationships, helping to allay lingering resentments and distrust, and create a good foundation for the future. One of the participants, a director of Partnerships First, said at the end of the workshop: 'That was the most productive meeting we've had in three years of partnering – it focused on the issues that make a difference', and that evidently expressed the common view.

Reviewing the hindsight process itself, Amicus concluded that:

- It is important to involve relevant people at all levels: project and site staff can benefit just as much from structured learning as top management. Involving project staff can be particularly valuable in a partnered environment because it encourages them to buy into the partnering ethos – and the best management intentions can be undermined if they do not.

- Workshops need to be skilfully facilitated – and facilitation is a skill that needs to be learned.
- Foresight reviews should be held as well, to ensure that best use is made of lessons learned in hindsight.
- A learning culture needs to be supported by positive encouragement to staff to be open, both in seeking support and resources, and in communicating progress and issues to be addressed; it does not happen automatically.

The Amphion Quality and Procurement Group subsequently decided to recommend a hindsight review as a key project milestone in all Amphion projects.

BAA

BAA is the largest international airport operator in the world, and one of the UK's biggest construction clients. In 2001 it owned Heathrow, Gatwick, Stansted, Glasgow, Edinburgh, Aberdeen and Southampton in the UK and had management contracts or stakes in 13 airports overseas, providing facilities for nearly 200 million passengers a year. BAA Property was responsible for developing and managing a £2 billion portfolio of support accommodation for the airlines using its airports, including offices, lounges, hangars, catering bases, cargo warehouses, hotels and check-in desks, making it one of the largest commercial property owners in the UK. Overall, BAA was spending around £1.5 million a day on new build, adaptation, refurbishment and maintenance projects, so it is not surprising that it put a strong emphasis on value for money. It was always looking for worthwhile innovation in construction management, and in recent years it had made sweeping changes in the way it handled construction projects. However, there was still a gap between individual learning and corporate initiatives, and no structured process to support corporate learning.

BAA's Fit Out Team took a local initiative to set up a simple, structured process designed to promote continuous learning and knowledge-sharing. This envisaged each project routinely including:

- One-hour meetings of managers and key professionals at the beginning of the definition, design, manufacture and assembly stages to feed in lessons from previous experience, and

- Two-hour reviews by the same staff at the end of each of these stages, followed by short collation exercises to extract key messages and record them on a spreadsheet.

However, the member of staff behind the proposal moved on, and it was not implemented.

The Fit Out Team remained interested in structured learning, and they decided to try out hindsight review, using it to look back at a £10 million project to develop a new check-in area in Gatwick South Terminal. This was created by building a new space to accommodate a ramp linking underused space to the existing passenger circulation area. The work involved designing and constructing the new space and ramp, and installing check-in desks and baggage-handling conveyors – all to a tight timetable and with the existing circulation area remaining in everyday use. After a feasibility study, work started on site in late 2000, and the new facility was brought into use in time for the 2001 Easter rush. The project was ultimately successful, but it had a chequered history of delay and cost increases, and it was an obvious opportunity to learn lessons with real business value. BAA's review took the form of:

- A desk review of project records
- Individual interviews (rather than a workshop), to give time for deep probing and encourage frank dialogue
- A post-review workshop to disseminate the lessons learned to a team involved in a similar, subsequent project.

The project record review provided a concise factual record of what happened, pinpointed the events most likely to be worth probing, and provided a basis for planning the interviews in detail and tailoring them to the individual interviewees. Each interview lasted between half an hour and an hour, ensuring that each interviewee's point of view was fully explored, and giving time for issues to be followed through. The combination of well-prepared interview scripts, good briefing of interviewees, adequate time, and privacy worked well. All the interviewees spoke remarkably freely, and the process revealed new insights into why things had happened and how difficulties could be avoided in future projects. It showed, for example, that pressures to bring completion deadlines forward without regard for what is physically achievable should be resisted more strongly. The gains from earlier handover are illusory, because it is more difficult, takes longer and costs more to complete construction work when an area is open to the public and in use than it would behind shutters.

The Fit Out Team later carried out a workshop-based hindsight review on another project to help them gauge the relative merits of interviews and workshops.

How it worked

The Fit Out Team were well satisfied with the outcome. The principle of structured learning had been accepted in the company before the trial started, but learning practice was inconsistent. There were local resource and cultural issues to address, and a lack of the tools that people need to make learning work. Practical experience with hindsight helped crystallise thinking on the way forward, and provided all the key tools.

It would not be cost-effective to carry out hindsight reviews as thoroughly as the first one after every project – BAA was carrying out around 500 construction projects a year at Heathrow alone – but team managers judged that they would be well worthwhile in projects that were high-value, were likely to be repeated, and offered clear opportunities to learn lessons. They planned to develop a more systematic approach to dissemination to extract even more value from them. The only significant change they expected to make to the process used in the pilot was to carry out reviews sooner after project completion, before memories and interest start to fade.

BAA took several lessons from the pilot hindsight exercise:

- It is important to make sure the staff responsible for the learning process have time to do it properly. Cutting corners saves very little, and can lose much more in lessons missed.
- Business value comes from applying lessons learned, not from collecting information. Workshops and interviews are only the beginning of the process; lessons need to be extracted from the records, and disseminated to the right audience.
- Structured learning needs strong support from the top, and the culture needs to be reinforced with appropriate incentives.
- Hindsight reviews should be carried out immediately after project completion, before memories and interest fade.

BP/Bovis Global Alliance

BP established the Global Alliance with Bovis Lend Lease in 1996 to reduce the cost of building and operating retail petrol stations across 11 European countries, and in the USA. An innovative contractual framework was used to align the business objectives of the two partners: Bovis Lend Lease's remuneration each year was linked to the Alliance's performance in areas such as health and safety, cost, time and quality, and maintenance, measured against agreed benchmarks. Personal incentives for UK employees were linked to pan-European weighted average performance. Not surprisingly, the whole culture of the Alliance emphasises teamwork, learning from experience and sharing knowledge across national boundaries as the keys to success.

The Alliance was able to build on Bovis Lend Lease's established approach to knowledge management. This sees it as the result of numerous daily acts of searching for better ways of doing things, rather than as a separate business activity. By the time of the case study (2002), the Alliance had run a successful 'Lessons Learned' initiative for several years, and Bovis's facilitated knowledge-sharing system *ikonnect* already gave all its staff access to lessons learned and expertise worldwide.

The performance benchmark agreed for the UK team in 2002 was to cut the total cost of a standard service station by more than 25% and save two weeks in construction time, while maintaining service levels. They were expected to:

- Make capital expenditure savings that could be replicated across the network in the UK and other countries
- Use new thinking to deliver BP's retail services more efficiently and effectively.

With savings of a similar magnitude already achieved in a previous round of improvements, this required major rethinking of the assumptions and functional requirements behind petrol station design, and they approached it using a combination of an extended foresight workshop and standard value management techniques.

A multidisciplinary team of eight was formed with representatives from the design and innovation, project management, commercial and procurement teams, together with selected external suppliers and contractors. Further knowledge and experience was brought in on an ad hoc basis from Bovis Lend Lease. Four half-day working sessions, one to two weeks apart and facilitated by the team leader, worked through three main phases:

- Understanding the factors affecting the achievement of the target, and identifying focus areas where improvement appeared most likely to be possible
- Brainstorming ideas and alternatives, using experience from previous projects
- Analysis and evaluation of proposals using preset evaluation criteria.

Phase 1: Understanding the challenge and identifying focus areas

In the first session the team reviewed information on cost and performance from historical projects, both to give them fresh insights into BP's functional requirements for service stations and to enable them to develop criteria for evaluating new proposals.

They evaluated every focus area and building element to find those that had both the potential to make a big impact on cost and scope for new kinds of solution. The easy cost-saving measures had been adopted in earlier years, so the team looked at whole systems, examining (for example) the design, source and installation of the whole electric system.

They designed their evaluation criteria to reflect the elements in the Alliance's scorecard – capital cost, life cycle efficiency, safety, and ease of installation and maintenance.

Phase 2: Functional analysis and brainstorming alternatives

There were two sessions in this stage. In the first, the team investigated the basic functionality required by each of the chosen areas and elements, focusing on what they wanted it to do rather than dwelling on its physical nature. This made space for creativity, enabling the discussion to range widely across functional solutions they had used or tried before. In the second session they brainstormed alternatives to current systems, looking for solutions that had worked before and would do what was required. Team members selected ideas they each wanted to pursue and develop further from the brainstormed list, according to their own areas of speciality.

Despite the division into two sessions, the identification of element functionality and brainstorming of ideas was actually iterative. Much of the learning arose from exchanges of tacit knowledge during this phase.

Phase 3: Proposal analysis and evaluation

In the last session each team member presented a draft proposal showing how their chosen solution could be implemented, and how it would pass the evaluation criteria.

How it worked

The Global Alliance's approach to foresight was designed to provide three key factors to stimulate creativity and facilitate the development of good solutions:

- A concentration of knowledge, information and experience, through bringing in people with experience of similar projects and understanding of what works
- Focus, to channel the available energy and resources into the areas that offered the greatest scope for improvement
- A stress on functionality rather than form, to create space for creativity and bring objectivity to the examination and evaluation of ideas.

It supported these with committed leadership, incentives for knowledge-sharing, and a clear aim.

The 2002 workshop more than achieved its objective: the evaluation suggested that the design finally adopted should reduce costs by 29–33%, beating the 25% target by a substantial margin. Overall, BP has estimated that the Alliance saved it around $800 million in its first 8 years, while improving quality and safety, and it credits good knowledge management practices with making important contributions to this.

Buro Happold

Consulting engineers Buro Happold used a hindsight exercise to learn lessons from their role in the first phase of a £21 million project to replace the Natural History Museum's 1920s zoology building. This started with site masterplanning in 1994; Buro Happold was appointed as structural, services and fire engineers some time later, and Phase One of the new Darwin Centre was opened to the public in 2002. The project involved several engineering challenges, not least the safe housing of a large amount of potentially explosive spirit in fragile containers: in addition to laboratories and offices, the new building accommodates 22 million specimens stored at low temperature in 450 000 jars of alcohol. Many of these have historical value as the first of their kind – including some collected by Darwin himself.

Buro Happold had recognised for some time that project teams learn many potentially valuable lessons in the course of their work, but had no effective way to capture them or disseminate them around the practice. Procedures require a close-out meeting at the end of every project, but these rarely happened, and even when they did their learning value was small. Searching for a way to make learning work better and share lessons learned more effectively, Buro Happold decided to try a hindsight review process based on:

- A desk review of project records.
- Individual interviews to capture tacit knowledge, because staff were already thought to be 'workshopped out', and it would be difficult to gather the Darwin Centre team together again for a group event (especially as some had since left the practice). Buro Happold judged that saving the time it would take them all to travel to a workshop would more than pay for the interviewer's time.
- A documentary report in the learning history format developed at MIT to disseminate lessons learned and make a durable record that would be a valuable addition to a new project database they were developing.

The review of project records revealed a number of problems that had arisen in the project, and other issues that promised to repay particular attention. It also gave the interviewer the background information needed to conduct the interviews intelligently and flexibly, avoiding the limitations of a rigid script.

Eight people were interviewed, including representatives of all the three engineering disciplines involved in the project. Those who had left the practice were interviewed at their new places of work. Nobody was interviewed from the client side, the other design practices involved in the project, or the contractors, partly because of uncertainty about their willingness to take part in the process and partly because a number of clients were being interviewed in a separate exercise. The project principal was interviewed first to get a comprehensive overview of the project, and to check and extend the list of significant issues identified in the desk review. All interviewees were briefed in advance on the objectives and format of the interviews, and reassured that they would not be used in personal assessments, and that quotations in the report would be anonymous. All the interviews were tape-recorded to avoid splitting the interviewer's attention between questioning and note-taking, and to allow verbatim transcripts to be produced.

The interviews were structured broadly on the What happened? – Why did it happen? – How can we do better? model. Although there had been some initial resistance to setting time aside, everyone spoke freely and openly once interviews had started. Each took between an hour and an hour and a half, several going on beyond the planned time. After the interviews were completed, the tapes were transcribed and analysed alongside the project records to identify lessons and ideas for improving practice that were likely to be valuable in Buro Happold's future work.

The results were documented in learning history format. The main features of this are:

- A basic division into 'chapters' dealing with particular episodes and issues – in Buro Happold's case, the main stages of the project from appointment, briefing, scheme and detailed design to implementation, together with a number of cross-cutting themes such as production information, programme and relationships.
- A secondary division into 'segments' focusing on particular dilemmas, questions or anecdotes, such as (in the chapter on briefing) client input and client priorities. Each segment begins with a short prologue that explains what it is about, and summarises the main facts and events that are relevant.
- The main text beneath the prologues, in a two-column format, with distilled lessons learned on the left juxtaposed

with verbatim quotations on the right to give them life and make them memorable. The quotations are attributed only by profession and generic job title, not by name.

The learning history was circulated to senior staff and later incorporated into a new project database.

How it worked

Reviewing their version of hindsight, Buro Happold concluded that it had worked well and represented a real step forward from conventional close-out reviews. In particular:

- A deliberate, structured approach to learning is much more effective than conventional project review meetings in which learning is secondary to management.
- Interviews need cost no more than workshops, and they can be a more flexible and practical way of gathering information when people feel themselves to be very busy and are geographically dispersed. This makes them more generally suitable for Buro Happold's circumstances than workshops. On the other hand, they lose the benefits of interaction, which can be vital in revealing team knowledge – knowledge that the team possesses collectively, but which is spread around in pieces that mean nothing until they are put together.
- The learning history format makes for a vibrant record, which helps readers absorb and remember. It is worthwhile for larger projects, but too labour-intensive for small ones.

Lattice Property

Lattice Property[1] was set up (as British Gas Properties) in 1994 to take over a large portfolio of property formerly belonging to British Gas, including some 1000 former gasworks ranging in size from 300 acres to half an acre, from Scotland to Cornwall. Its mission was to reclaim the land – 7000 acres of it potentially contaminated – and return it to beneficial use. By 2002 Lattice had already cleaned up hundreds of acres of brownfield land, which have since been redeveloped as industrial units, leisure facilities, homes, schools, shops and open space, and it was restoring around 70 sites a year.

The technical and managerial problems facing Lattice in 2002 were surprisingly varied, and the regulatory environment in which it worked was becoming increasingly stringent. Contaminated sites have to be

---

[1] Now National Grid Property.

treated, often to a considerable depth. The dirtiest and most danger-
ous material is excavated and sent to licensed tips. Cleaner material,
such as building rubble, can often be left *in situ*, capped off with
compacted clean clay and soil. In between these extremes there is a
large volume of moderately contaminated soil that could be dealt with
in two ways. In the past it has normally been sent to tips, but it can,
in principle, be 'bioremediated' using aerobic bacteria, which convert
hydrocarbon wastes to non-toxic carbon dioxide and water. In this
process the excavated spoil is heaped into long mounds, primed with
bacteria and nutrients, turned over repeatedly for 4–10 weeks, and
then put back. In 2002 this was still a relatively expensive and unproven
option, but rising landfill tax and a falling number of tips licensed to
accept contaminated material were making it an increasingly attractive
option.

One of Lattice's first experiences of bioremediation was in the
second phase of a project to regenerate a large former gasworks site
in Portsmouth. A first phase had been completed several years previously and the land redeveloped, but the larger part of the site had
been left untouched because investigation showed that it would cost
an uneconomic £6–7 million to clean up using conventional methods.
A new costing in 2000/2001 suggested that bioremediation might
offer an economically viable as well as environmentally superior way
forward, but it was not yet widely accepted by local authority environmental officers. Portsmouth City Council were prepared to allow
it only under very strict conditions: they said, for example, that they
would stop the work if they received more than two complaints about
smell. Lattice nevertheless decided to go ahead and make Portsmouth
a trial of large-scale bioremediation. Despite several difficulties –
including exceptionally wet weather (which complicated the bioremediation process) and the discovery of unexpectedly larger volumes of
heavily contaminated material – the project was completed successfully and on time. Lattice was keen to learn all it could from the experience, especially as it had been conducted under an exceptionally
rigorous regulatory regime. It expected many of the lessons to carry
over directly to future projects.

With no previous experience of structured learning, Lattice decided
to carry out a workshop-based hindsight review. This started in the
last few weeks of the project with a desk review of project records to
identify key stages, and the review workshop followed shortly after
project completion. This was attended by seven senior and middle
managers from all the main organisations involved in the project, and
was scheduled to take two and a half hours, but overran by about half
an hour. The discussion followed the classical sequence and time plan
– about 25% establishing what happened, 25% discussing why it happened, and 50% working out how to do better in future – using the

chronology from the desk review as a framework. In the event there was some blurring of boundaries between the three parts of the workshop, because the workshop leader found it difficult to stop discussion of what had happened (and why) leading immediately into suggestions for doing better. The discussion was open, animated and constructive throughout: participants had been briefed on the ground rules before the workshop, and had no difficulty following them.

The exercise provided new insights into the factors that lay behind the success of the Portsmouth project, and identified others that had caused avoidable problems. It showed, for example, that having a team with the right combination of expertise and good working relationships had been crucial in allowing the project to overcome difficulties; that it pays big dividends to investigate sites thoroughly before planning projects and to prepare paperwork for regulators before work starts; that it is best to use the same contractor and plant for excavating and for turning windrows to avoid conflicting aims; and that efficient operation of the batch processes involved in bioremediation requires site managers to plan continuously 2–3 weeks ahead.

Overall, around 20 significant lessons about bioremediation and project management emerged. To make these available to all the managers responsible for remediation projects, Lattice encapsulated all the lessons in a three-page guidance note detailing issues that should be taken into account in the key stages of site investigation review, team assembly, detailed planning, and obtaining regulator approval.

How it worked

Lattice found that the workshop process worked well, and decided to adopt it as part of the comprehensive approach to a knowledge management system that it was developing. The structured, facilitated workshop was much more productive of lessons with real business value than the conventional project management meetings it had relied on in the past. It also helped to ensure that lessons learned on the job – as many are when relationships between client and contractors are good – were recognised and disseminated. The process promised to be particularly valuable when, as in the Portsmouth project, novel techniques are being used and overall success tended to leave problems unexamined.

Lattice concluded that:

- It is important to plan adequate time for workshops: the benefits easily outweigh the small extra cost.
- To make best use of available time, and make the workshop leader's job easier, it is helpful to pre-filter issues to focus attention on the most important, and to have the basic

chronology of the project visible throughout the workshop, perhaps on a flip chart. A well-structured discussion is also easier to record in a form that can easily be developed into a tidy document to disseminate lessons learned.

- There is a tendency for discussion of what happened and why to spill over into discussion of how to do better. Within limits this does not matter, but the workshop leader should not allow it to compromise coverage of the whole project history and all the important issues.
- Workshops have an inevitable bias towards the circumstances of the project being reviewed. Workshop leaders need to be aware of this and steer discussion of how to do better towards circumstances likely to arise in future projects.
- A structured approach may not make learning much more effective within small teams, but it pays big dividends by enabling lessons to be captured and disseminated throughout an organisation.

Epilogue

Chapter Twenty-Six
Where Next for
Knowledge Management?

The development of competence has been described as a four-stage journey:

1. *Unconscious incompetence*, when you don't even realise that an area of expertise exists, or you know it exists but don't think it is relevant to you
2. *Conscious incompetence*, when you become aware that it exists and it would be worth acquiring
3. *Conscious competence*, when you have acquired some expertise and you can use it reliably, but it takes conscious effort
4. *Unconscious competence*, when it becomes second nature, fully integrated into the way you think and behave.

Before 1990 few companies had ever given any coherent consideration to their knowledge. They all took skills into account in recruitment and promotion, most set up training programmes, and many attached great importance to tradable assets such as copyright, patents and brands, but very few managers thought consciously about how people learn from practical experience and how knowledge flows around. That was just left to happenstance, with results varying widely in speed and success. Even fewer thought about the possibility or the benefits of improving learning or knowledge-sharing. Nearly everyone was at stage 1, where ignorance is bliss. Today, nearly all organisations have reached stage 2, many are moving tentatively into stage 3, and a few have reached stage 4. For most, the journey is proving to be bumpy, halting and slow, as the case studies show – and these are all taken from organisations that are highly successful in their different ways, with outstanding management teams. It looks like being many years before knowledge becomes a background factor in all management thinking, as finance is today.

But even the unconsciously competent are not necessarily expert, and in most fields technological development and increasing theoretical understanding are continuously raising the bar. What can we expect to happen to the theory and practice of knowledge management? These are my speculations.

Web 2.0

One influence that is already having a visible effect is the evolution of *Web 2.0*, the gradual change from a web of corporate publishers and personal readers to a social commons where more and more of the content is contributed by the readers themselves, and where they interact directly with each other.

The creators of the original Web (1.0) included user generation of content and collaboration among its goals, but it has taken software tools and architectures such as Java, wikis, blogs, mash-ups and tag clouds to make them a reality. Now that they are, we can exploit in private corporate webs the technical tools and – equally importantly – the attitudes and skills that users are developing as they look for contacts in sites such as Facebook and LinkedIn, share their favourite

Web 1.0	Web 2.0
Content from corporation owning site	Content from corporation owning site *and* aggregated from other sites *and* from users
Functionality provided by site owner	Functionality provided by site owner *and* from other sources
Structure fixed by site owner	Structure fixed by site owner *or* created by users
Links to pages in same and other sites	Links to pages *and data* in same and other sites
Content changes slowly	Content changes rapidly
Publishing	Publishing *and* collaboration
Lecture	Lecture *and* conversation
Information source for users	Information source *and* communications tool for users
Few originators, many readers	Many originators *and* readers
Users unaware of each other	Users interact with each other
Impersonal	Social
Connects top-down	Connects down, up and across
Corporate authority	Caveat emptor

Table 26.1 Evolution from Web 1.0 to Web 2.0.

websites though Digg or Reddit, post holiday snaps in Flickr, contribute to Wikipedia, review their purchases on Amazon and suppliers on Shopping.com, and rate pages and vote in online polls. Simple wiki knowledge bases, networking directories and blogs where users write and sign their own contributions are the first examples of this. They are already used in many companies, but fear of quality problems, potential risks and general loss of control is holding them back in others: this will surely pass. It will take time to discover the full potential of Web 2.0, but even in its current forms it is undoubtedly a major step forward in the technology of knowledge management.

The semantic web

The rich interactivity of Web 2.0 evolved from elements such as forums and feedback boxes that have been in use for a long time. Similarly, Web 2.0 contains the seeds of what is sometimes called the 'semantic web' in price comparison and news sites that aggregate data from a variety of publishers. The potential for adding value to information by doing this is obvious. As humans, we are used to picking up information from numerous sources, putting it together with our own mental models, and making new knowledge out of it. We are very good at this: we ignore surrounding irrelevance such as adverts and journalistic titles designed to catch the eye, extract what we need, translate between different terminologies, guess when we meet ambiguities and gaps, and make it all grist to our mill. All sorts of possibilities would open up if computers, with their almost instantaneous 'reading' and 'thinking', could do the same . . . but they can't. We cope so easily with the difficulties that we hardly notice most of them, but computers are so ignorant and stupid that they would be stumped by every one. As far as they are concerned, most information is irretrievably stuck in the silos of individual web pages and documents. The semantic web aims to change that.

Price comparison sites can aggregate information usefully from numerous sources because it is highly structured: products all have predictable characteristics such as manufacturers, model numbers and prices. The semantic web uses a variety of tools to 'understand' text and numbers so that it can do the same with material that is less structured and more varied. Tim Berners-Lee, the inventor of the Web (and a leading light in the development of the semantic web), cites the search for new medical treatments as an area where this could have a dramatic impact. At the moment researchers have to try to keep up to date with already vast and rapidly expanding knowledge about disease symptoms, cell biology, organ function, the pharmacological effects of known chemical compounds, genetics, experimental techniques and more. This is all scattered between academic papers, trials databases, textbooks, slides, manufacturers' literature and other locations, and only a human can make sense of it and see new con-nections that might lead to progress. If computers could do that – even relatively unintelligently – they could explore possibilities far faster than any human can, point out connections that nobody has ever thought of, and enable researchers to focus their efforts in the most promising areas.

Large corporations such as IBM, Nokia and Oracle are already experimenting with systems to do this in restricted domains, and com-mercial software tools are under development. When practical and affordable ways are developed to let any organisation do the same with the information in their project documentation, knowledge bases, correspondence, accounts and databases, and the semantic web becomes a reality instead of just a seductive idea, knowledge man-agement will take another important step forward.

Developments in psychology and the science of human relations

The democratisation of the workplace and the flattening of corporate structures have already weakened the comforting (if largely illusory) certainties of top-down authority, and Web 2.0 looks like eroding them further. Knowledge management has so far been largely empirical and pragmatic, but it will need increasingly to be informed by a more sophisticated understanding of psychology and human relations.

There is a wealth of material to draw on: research results on topics such as game theory, emotional intelligence, social influences and tipping points, behavioural economics, motivation, trust, reciprocity and altruism all have relevant insights to offer, and their emergence from the obscurity of research papers into the popular science and management literature has made them accessible. However, this has so far had little perceptible effect on the design of networking tools or wiki knowledge bases, on the organisation of communities of prac-tice, or on other aspects of knowledge management or the manage-ment of change. Only workspace design has benefited significantly. It seems to me that research on psychology and human relations has the potential to reveal more of the reasons why some implementa-tions of knowledge tools and techniques disappoint, to help design better versions, to show how to manage initiatives more effectively, and ultimately to lift knowledge-sharing (and to a lesser extent learn-ing) onto a new level.

Insights from neuroscience

The development of MRI imaging and other tools for looking into the working of the human brain at levels down to the individual neuron has hugely accelerated progress in understanding perception, memory and learning, and together with psychological research it is leading to a much deeper understanding of the basis of expertise than we had only a few years ago. I believe that applying this to knowledge management will show organisations how to become much better at developing the expertise of their staff, and significantly accelerate their progress from raw graduates into capable professionals, and (for some) from capable professionals into outstanding practitioners.

Empirical improvements

Technical developments and insights from research in the human sciences look like producing the biggest improvements in knowledge management over the next few years, but there is still useful scope for empirical development. By definition this will come from experience, and that makes it unpredictable. However, three areas seem to me to be evidently promising:

- The development of a more differentiated approach to knowledge, and triage techniques for deciding how best to communicate different kinds of information and expertise
- A better understanding of the relationships between tools, processes, and the specific circumstances and needs of different kinds of work, and finally
- Better understanding of the particular needs of professionals and cross-fertilisation between the approaches to knowledge management used by different professions.

Together, Web 2.0, the semantic web, insights from research in psychology, human relations and brain science, and empirical improvements seem to me to have the potential to revolutionise knowledge management and make it much more scientific, certain and rewarding than it generally is today.

Further Reading

Conscious that most readers of this book are likely to be busy working professionals, I have tried throughout to avoid straying too far from the topics and ideas that I think everyone with a serious practical interest in managing organisational knowledge should be acquainted with, and to avoid delving deeper than necessary into them. Inevitably, that means that I have had to leave much out. The books and journal articles cited here can fill most of the gaps, and they often provide alternative ways of thinking about the issues as well. There are many others; these are the ones I have found most informative and inspirational.

With one exception (the *Blackwell Handbook of Organizational Learning and Knowledge Management*), they are all written for a business or general audience, and most are highly readable. They are not academic books, although many of the authors are academics. Knowledge management is, after all, all about making things happen in the real world.

They are listed under broad topic headings, but the distinctions should not be taken too literally; there are some appreciable overlaps, particularly between learning from experience and knowledge-sharing.

Business strategy, leadership and managing change

The Discipline of Market Leaders: Choose your customers, narrow your focus, dominate your market, Michael Treacy and Fred Wiersema, Basic Books, 1995.

Understanding Organisations, Charles B. Handy, 3rd edn, Penguin Books, 1985. A broad overview of key concepts including motivation, leadership, power and influence, the working of groups, organisational culture, and developing people. Full of illuminating insights, like all Charles Handy's books.

Making it Happen: Reflections on leadership, John Harvey-Jones, Profile Books, 2003.

Leading change: why transformation efforts fail. John P. Kotter, *Harvard Business Review*, 1995, March–April, pp. 59–67.

Leading Change, John P. Kotter, Harvard Business School Press, 1996. Kotter's classic work on change management, expanding on the ideas summarised in his previous *Harvard Business Review* article.

The Heart of Change: Real life stories of how people change their organizations, John P. Kotter and Dan S. Cohen, Harvard Business School Press, 2002. A follow-up to *Leading Change*.

The Knowing–Doing Gap: How smart companies turn knowledge into action, Jeffrey Pfeffer and Robert I. Sutton, Harvard Business School Press, 2000. Discusses the problem that the top management in many companies know what they should do but fail to do it.

Knowledge management

The business case

Intellectual Capital: The new wealth of organizations, Thomas A. Stewart, Nicholas Brealey Publishing, 1997.

Knowledge Unplugged: The McKinsey & Company global survey on knowledge management, Jurgen Kluge, Wolfram Stein and Thomas Licht, Palgrave, 2001. Persuasive evidence for the business benefits of KM.

Unleashing the power of learning: an interview with British Petroleum's John Browne. *Harvard Business Review*, 1999, March, pp. 147–168, or reprint number 97507.

Rethinking Construction: The report of the Construction Task Force to the Deputy Prime Minister on the scope for improving the quality and efficiency of UK construction, Department of the Environment, Transport and the Regions, 1998.

Principles and underlying mechanisms

What's your strategy for managing knowledge? Morten T. Hansen, Nitin Nohria and Thomas Tierney, *Harvard Business Review*, 1999, March–April, or reprint number 99206.

The Knowledge-Creating Company: How Japanese companies create the dynamics of innovation, Ikujiro Nonaka and Hirotaka Takeuchi, Oxford University Press, 1995.

Enabling Knowledge Creation: How to unlock the mystery of tacit knowledge and release the power of innovation, Georg von Krogh, Kazuo Ichijo and Ikujiro Nonaka, Oxford University Press, 2000. One of the classic texts, interesting as general background. It is concerned more with principles than with specific techniques, and with large companies and mixed workforces.

Working Knowledge: How organizations manage what they know, Thomas A. Davenport and Laurence Prusak, Harvard Business School Press, 2000. Another classic text.

Sticky Knowledge: Barriers to knowing in the firm, Gabriel Szulanski, Sage, 2003. Reports the results of Szulanski's detailed investigation of the causes of 'stickiness' – essentially failure to share and replicate good practice – in a number of major US corporations: a research report, not a practical guide.

The Hidden Power of Social Networks: Understanding how work really gets done in organisations, Rob Cross and Andrew Parker, Harvard Business School Press, 2004. Includes a detailed, practical discussion of social network analysis.

Knowledge Management: A state of the art guide, Paul Gamble and John Blackwell, Kogan Page, 2001. A good general discussion of knowledge management, wide-ranging, coherent and generally down to earth; more about principles than practicalities.

The Blackwell Handbook of Organizational Learning and Knowledge Management, Mark Easterby-Smith and Marjorie A. Lyles, Blackwell, 2005. A highly detailed academic review of research in organisational learning, the 'learning organisation' and knowledge management from their origins in the 1960s to around 2000. Most of the sources cited are from the mid 1990s or earlier, and predate widespread use of knowledge management in industry and the practical experience it has provided, so this is interesting principally as historical background.

Learning from experience

Learning in Action: A guide to putting the learning organisation to work, David A. Garvin, Harvard Business School Press, 2000. A good general discussion of the learning process, specific tools and techniques, and the rationale for them.

Hope is Not a Method: What business leaders can learn from America's army, Gordon R. Sullivan and Michael V. Harper, Broadway, 1997. The story of the US Army's transformation since the end of the Cold War, including its development and use of After Action Reviews.

Car Launch: The human side of managing change, George Roth and Art Kleiner, Oxford University Press, 2000. A detailed and readable case study on MIT Sloan School's 'learning histories' technique and its use to learn lessons from the development and launch of a new car model.

Oil Change: Perspectives on corporate transformation, George Roth and Art Kleiner, Oxford University Press, 2000. Another detailed case study on learning histories, this time used in the context of a corporate change programme in a major international oil company.

The Fifth Discipline: The art and practice of the learning organization, Peter Senge, Doubleday, 1990.

The Fifth Discipline Fieldbook: Strategies and tools for building a learning organization, Peter Senge, Art Kleiner, Charlotte Roberts, Richard Ross and Bryan Smith, Nicholas Brealey Publishing, 1994.

The Dance of Change: The challenges of sustaining momentum in learning organizations, Peter Senge, Art Kleiner, Charlotte Roberts, Richard Ross, George Roth and Bryan Smith, Nicholas Brealey Publishing, 1999.

Sharing knowledge

If Only We Knew What We Know: The transfer of internal knowledge and best practice, Carla O'Dell and C. Jackson Grayson, The Free Press,

1998. A non-academic overview of knowledge-sharing from the President and Chairman of the American Productivity and Quality Center. Draws extensively on ideas from *The Discipline of Market Leaders* and *Sticky Knowledge*.

Learning to Fly: Practical lessons from one of the world's leading knowledge companies, Chris Collison and Geoff Parcell, Capstone, 2001. A detailed and down-to-earth description of BP Amoco's KM systems and the thinking behind them, by two of their creators.

Common Knowledge: How companies thrive by sharing what they know, Nancy M. Dixon, Harvard Business School Press, 2000. Identifies and discusses five different situations: when the same team repeats the same task in a new context ('serial transfer'); transferring knowledge from one team to another doing a similar job in a similar context ('near transfer'); transferring tacit knowledge about 'non-routine' tasks between teams ('far transfer'); sharing very complex knowledge between teams ('strategic transfer'); and transferring explicit knowledge about an uncommon, specialist task ('expert transfer').

Deep Smarts: How to cultivate and transfer enduring business wisdom, Dorothy Leonard and Walter Swap, Harvard Business School Press, 2005. Discusses the importance of 'deep smarts: a potent form of experience-based wisdom that drives both organizational competitiveness and personal success', and how to cultivate and exploit it.

The Springboard: How storytelling ignites action in knowledge-era organisations, Stephen Denning, Butterworth-Heinemann, 2001. Tells the story of how Stephen Denning lit on storytelling as a knowledge management tool while working at the World Bank, and discusses the principles and practice in detail.

Cultivating Communities of Practice: A guide to managing knowledge, Etienne Wenger, Richard McDermott and William Snyder, Harvard Business School Press, 2002. Comprehensive, authoritative and readable, albeit with some bias towards managing rather than simply encouraging CoPs – an approach most authors advise against.

Continuity Management: Preserving corporate knowledge and productivity when employees leave, Hamilton Beazley, Jeremiah Boenisch and David Harden, Wiley, 2002.

Making a Success of Co-located Design and Construction Teams, The Business Round Table, 2006.

Psychology and the brain

The Human Mind – and how to make the most of it, Robert Winston, Bantam Books, 2004. A very readable basic overview.

Your Memory: A user's guide, Alan Baddeley, McGraw-Hill, 1982. Another very readable introduction.

How the Mind Works, Steven Pinker, Penguin Books, 1998. An excellent and much deeper book by one of the world's leading cognitive scientists.

The Making of Memory: From molecules to mind, Steven Rose, Bantam Books, 1993. More specialised but otherwise comparable with Pinker's book.

Making Up the Mind: How the brain creates our mental world, Chris Frith, Blackwell Publishing, 2007. An excellent, indispensable and up-to-date overview by a leading professor of neuropsychology.

A Mind of its Own: How your brain distorts and deceives, Cordelia Fine, Icon Books, 2007. A good, popular account of the messy realities of human thought.

How the Mind Forgets and Remembers: The seven sins of memory, Daniel L. Schacter, Souvenir Press, 2003.

Six Impossible Things Before Breakfast: The evolutionary origins of belief, Lewis Wolpert, Faber & Faber, 2007. Discusses how practical experience and the brain's search for causality shape our mental model of the world.

Blink: The power of thinking without thinking, Malcolm Gladwell, Penguin Books, 2006. Further insights into the way we think.

The Motivated Mind: How to get what you want from life, Raj Persaud, Bantam Press, 2005. Ostensibly a self-help book, but full of insights relevant to knowledge management (and management in general).

Freakonomics: A rogue economist explores the hidden side of everything, Steven D. Levitt and Stephen J. Dubner, Penguin Books, 2006. Insights into the ways in which motivation and self-interest affect behaviour, sometimes in ways that seem contrary to common sense.

Working With Emotional Intelligence, Daniel Goleman, Bloomsbury, 1998. The implications of emotional intelligence for organisations.

How People Learn: Brain, mind, experience and school, John B. Bransford, Ann L. Brown. and Rodney R. Cocking (eds), Committee on Developments in the Science of Learning, National Academy Press, 2000. An excellent, non-academic synopsis of the science of learning and its implications for teaching and learning.

Critical Mass: How one thing leads to another, Philip Ball, Arrow Books, 2005. A discussion of the interplay and dynamics of culture, customs, institutions, cooperation and conflict.

The Tipping Point: How little things can make a big difference, Malcolm Gladwell, Abacus, 2001. Persuasive evidence that the devil really is in the detail.

Nudge: Improving decisions about health, wealth and happiness, Richard H. Thaler and Cass R. Sunstein, Yale University Press, 2008. Shows how subtle 'nudges' can be used to help people avoid the biases, blunders, temptations and social pressures that so often lead them to make poor choices. Thought-provoking reading for anyone trying to create a learning organisation.

Index